Semaphorins: Receptor and Intracellular Signaling Mechanisms

ADVANCES IN EXPERIMENTAL MEDICINE AND BIOLOGY

Editorial Board:
NATHAN BACK, *State University of New York at Buffalo*
IRUN R. COHEN, *The Weizmann Institute of Science*
ABEL LAJTHA, *N.S. Kline Institute for Psychiatric Research*
JOHN D. LAMBRIS, *University of Pennsylvania*
RODOLFO PAOLETTI, *University of Milan*

Recent Volumes in this Series

Volume 592
REGULATORY MECHANISMS OF STRIATED MUSCLE CONTRACTION
Edited by Setsuro Ebashi and Iwao Ohtsuki

Volume 593
MICROARRAY TECHNOLOGY AND CANCER GENE PROFILING
Edited by Simone Mocellin

Volume 594
MOLECULAR ASPECTS OF THE STRESS RESPONSE
Edited by Peter Csermely and Laszlo Vigh

Volume 595
THE MOLECULAR TARGETS AND THERAPEUTIC USES OF CURCUMIN IN HEALTH AND DISEASE
Edited by Bharat B. Aggarwal, Yung-Joon Surh and Shishir Shishodia

Volume 596
MECHANISMS OF LYMPHOCYTE ACTIVATION AND IMMUNE REGULATION XI
Edited by Sudhir Gupta, Frederick Alt, Max Cooper, Fritz Melchers and Klaus Rajewsky

Volume 597
TNF RECEPTOR ASSOCIATED FACTORS (TRAFs)
Edited by Hao Wu

Volume 598
INNATE IMMUNITY
Edited by John D. Lambris

Volume 599
OXYGEN TRANSPORT TO TISSUE XXVIII
Edited by David Maguire, Duane F. Bruley and David K. Harrison

Volume 600
SEMAPHORINS: RECEPTOR AND INTRACELLULAR SIGNALING MECHANISMS
Edited by R. Jeroen Pasterkamp

A Continuation Order Plan is available for this series. A continuation order will bring delivery of each new volume immediately upon publication. Volumes are billed only upon actual shipment. For further information please contact the publisher.

Semaphorins: Receptor and Intracellular Signaling Mechanisms

Edited by
R. Jeroen Pasterkamp
Department of Pharmacology and Anatomy, Rudolf Magnus Institute of Neuroscience, University Medical Center Utrecht, Utrecht, The Netherlands

Springer Science+Business Media
Landes Bioscience

Springer Science+Business Media
Landes Bioscience

Copyright ©2007 Landes Bioscience and Springer Science+Business Media

All rights reserved.
No part of this book may be reproduced or transmitted in any form or by any means, electronic or mechanical, including photocopy, recording, or any information storage and retrieval system, without permission in writing from the publisher, with the exception of any material supplied specifically for the purpose of being entered and executed on a computer system; for exclusive use by the Purchaser of the work.

Printed in the U.S.A.

Springer Science+Business Media, 233 Spring Street, New York, New York 10013, U.S.A.
http://www.springer.com

Please address all inquiries to the Publishers:
Landes Bioscience, 1002 West Avenue, 2nd Floor, Austin, Texas 78701, U.S.A.
Phone: 512/ 637 6050; FAX: 512/ 637 6079
http://www.landesbioscience.com

Semaphorins: Receptor and Intracellular Signaling Mechanisms edited by R. Jeroen Pasterkamp, Landes Bioscience / Springer Science+Business Media, LLC dual imprint / Springer series: Advances in Experimental Medicine and Biology

ISBN: 978-0-387-70955-0

While the authors, editors and publisher believe that drug selection and dosage and the specifications and usage of equipment and devices, as set forth in this book, are in accord with current recommendations and practice at the time of publication, they make no warranty, expressed or implied, with respect to material described in this book. In view of the ongoing research, equipment development, changes in governmental regulations and the rapid accumulation of information relating to the biomedical sciences, the reader is urged to carefully review and evaluate the information provided herein.

Library of Congress Cataloging-in-Publication Data

Semaphorins : receptor and intracellular signaling mechanisms / edited by R. Jeroen Pasterkamp.
　p. ; cm.
　ISBN-13: 978-0-387-70955-0
　1. Semaphorins. 2. Cell receptors. 3. Cellular signal transduction. I. Pasterkamp, R. Jeroen.
　[DNLM: 1. Intercellular Signaling Peptides and Proteins--physiology. 2. Semaphorins--physiology.
3. Receptors, Cell Surface--physiology. QU 55.2 S471 2007]
　QP552.S32S46 2007
　572'.696--dc22
　　　　　　　　　　　　　　2007011220

About the Editor...

R. JEROEN PASTERKAMP is an Assistant Professor at the Rudolf Magnus Institute of Neuroscience, University Medical Center Utrecht, Utrecht, The Netherlands. The focus of his laboratory is directed towards understanding the intracellular signaling events involved in the formation of neuronal connections. His research team concentrates on the developing mouse embryo using an integrated approach involving molecular biology, cell biology, in vivo functional proteomics, and mouse genetics. He received his Ph.D. from the Netherlands Institute for Neurosciences (Amsterdam, The Netherlands) and did his Postdoctoral at the Department of Neuroscience, Johns Hopkins University School of Medicine, Baltimore.

PREFACE

Since the identification of the first two semaphorins in the early 1990s, Sema-1a (Fasciclin IV) and Sema3A (collapsin), more than 25 semaphorin genes have been described. Although originally identified as repulsive guidance signals for extending axons, these secreted and membrane-associated glycoproteins are extremely pleiotropic, and many serve diverse roles unrelated to axon guidance. These include multiple distinct roles within a given biological system or tissue, including axon guidance, cell migration and neuronal apoptosis in the nervous system, and also parallel functions in seemingly disparate systems, such as cell migration in the nervous, cardiovascular and immune systems. Our knowledge of the cellular actions of semaphorin family members has advanced significantly over the past several years, and the receptors and intracellular signaling mechanisms that underlie semaphorin function are being unveiled at a rapid pace.

Although plexins are the predominant family of semaphorin receptors, multiple (co-)receptor proteins function in several semaphorin signaling events. A unifying principle that defines the function of high-affinity semaphorin receptors characterized to date is their multimeric character. Unrelated receptor proteins with distinct functions (e.g., ligand-binding, signal-transducing, modulatory) are assembled into large holoreceptor complexes to detect and respond to semaphorin proteins present in the extracellular space. There is a growing appreciation that the composition of a semaphorin receptor complex not only determines ligand specificity and sensitivity but also dictates the functional outcome of a ligand-receptor interaction. For example, a semaphorin receptor complex may trigger attractive or repulsive cell migration events in response to the same semaphorin ligand depending on the presence or absence of certain co-receptor proteins.

Recent semaphorin research is characterized by an impressive effort to decipher the intracellular signal transduction networks downstream of semaphorins and their receptors. An ever-increasing number of cytosolic signaling molecules is being implicated in semaphorin signaling, and common principles begin to emerge that underlie the molecular basis of semaphorin function. Activation of small GTPases, and phosphorylation of both receptors and intracellular effector proteins, are crucial for semaphorin-mediated effects. In addition, recent evidence suggests the involvement of less well-known modulatory mechanisms, such as redox signaling and local protein synthesis.

The chapters included in this book are intended to provide a representative survey of recent progress on research devoted to semaphorins with an emphasis on receptor and intracellular signaling mechanisms. The first four chapters address several of the key families of cytosolic signaling cues implicated in semaphorin signaling (including CRMPs, small GTPases, protein kinases, and MICALs). The following three chapters cover the intracellular and extracellular factors that modulate semaphorin signaling (various cytosolic cues, Ig superfamily cell adhesion molecules, and proteoglycans). Finally, the last four chapters review recent progress in our understanding of how semaphorin signaling pathways may contribute to the development and disease of specific biological systems.

It is likely that work on semaphorin function such as outlined in this book will continue to advance our understanding of key regulator influences on cellular morphology, and that these studies will serve as a model for the complex and elaborate cellular and molecular functions of all major families of guidance cues.

R. Jeroen Pasterkamp

PARTICIPANTS

Aminul Ahmed
MRC Centre for Developmental
    Neurobiology
King's College London
London
U.K.

Ahmad Bechara
CGMC UMR 5534
Université Claude Bernard
Villeurbanne
France

Laurence Boumsell
INSERM U659
Faculté de Médecine
Créteil
France

Andrea Casazza
Division of Molecular Oncology
Institute for Cancer Research
    and Treatment (IRCC)
University of Turin Medical School
Candiolo, Torino
Italy

Valérie Castellani
CGMC UMR 5534
Université Claude Bernard
Villeurbanne
France

Britta J. Eickholt
MRC Centre for Developmental
    Neurobiology
King's College London
London
U.K.

Julien Falk
CGMC UMR 5534
Université Claude Bernard
Villeurbanne
France

Pietro Fazzari
Division of Molecular Oncology
    of the Institute for Cancer Research
    and Treatment (IRCC)
University of Turin Medical School
Candiolo, Torino
Italy

Cynthia Healy
ATC, Beckman Coulter
Miami, Florida
U.S.A.

Ofra Kessler
Cancer and Vascular Biology
    Research Center
Rappaport Research Institute
    in the Medical Sciences
The Bruce Rappaport Faculty
    of Medicine, Technion
Israel Institute of Technology
Haifa
Israel

Hitoshi Kikutani
Department of Molecular Immunology
Research Institute for Microbial Diseases
Osaka University
Osaka
Japan

Participants

Sharon M. Kolk
Department of Pharmacology
  and Anatomy
Rudolf Magnus Institute
  of Neuroscience
University Medical Center Utrecht
Utrecht
The Netherlands

Tali Lange
Cancer and Vascular Biology
  Research Center
Rappaport Research Institute
  in the Medical Sciences
The Bruce Rappaport Faculty
  of Medicine, Technion
Israel Institute of Technology
Haifa
Israel

Guo-li Ming
Institute for Cell Engineering
Departments of Neurology
  and Neuroscience
Johns Hopkins University School
  of Medicine
Baltimore, Maryland
U.S.A.

Frédéric Moret
CGMC UMR 5534
Université Claude Bernard
Villeurbanne
France

Patrick Nasarre
CNRS-UMR 6187, IPBC
Faculté des Sciences de Poitiers
Poitiers
France

Gera Neufeld
Cancer and Vascular Biology
  Research Center
Rappaport Research Institute
  in the Medical Sciences
The Bruce Rappaport Faculty
  of Medicine, Technion
Israel Institute of Technology
Haifa
Israel

R. Jeroen Pasterkamp
Department of Pharmacology
  and Anatomy
Rudolf Magnus Institute of Neuroscience
University Medical Center Utrecht
Utrecht
The Netherlands

Vincent Potiron
CNRS-UMR 6187, IPBC
Faculté des Sciences de Poitiers
Poitiers
France

Andreas W. Püschel
Abteilung Molekularbiologie
Institut für Allgemeine Zoologie
  und Genetik
Westfälische Wilhelms-Universität
Münster
Germany

Joëlle Roche
CNRS-UMR 6187, IPBC
Faculté des Sciences de Poitiers
Poitiers
France

Eric F. Schmidt
Department of Neurology
Yale University School of Medicine
New Haven, Connecticut
U.S.A.

Sangwoo Shim
Institute for Cell Engineering
Departments of Neurology
  and Neuroscience
Johns Hopkins University School
  of Medicine
Baltimore, Maryland
U.S.A.

Stephen M. Strittmatter
Department of Neurology
Yale University School of Medicine
New Haven, Connecticut
U.S.A.

Luca Tamagnone
Institute for Cancer Research
 and Treatment (IRCC)
University of Turin Medical School
Candiolo, Torino
Italy

Toshihiko Toyofuku
Department of Cardiovascular Medicine
Osaka University Graduate School
 of Medicine
Osaka
Japan

Asya Varshavsky
Cancer and Vascular Biology
 Research Center
Rappaport Research Institute
 in the Medical Sciences
The Bruce Rappaport Faculty
 of Medicine, Technion
Israel Institute of Technology
Haifa
Israel

Joost Verhaagen
Neuroregeneration Laboratory
Netherlands Institute for Neuroscience
Amsterdam
The Netherlands

Joris de Wit
Department of Functional Genomics
Center for Neurogenomics
 and Cognitive Research
Vrije Universiteit Amsterdam
Amsterdam
The Netherlands

CONTENTS

PREFACE VI

1. THE CRMP FAMILY OF PROTEINS AND THEIR ROLE IN SEMA3A SIGNALING 1

Eric F. Schmidt and Stephen M. Strittmatter

Properties and Expression of CRMPs 1
Evidence For CRMPs in Sema3A Signaling 2
The Regulation of the Cytoskeleton by CRMPs 5
CRMP, Endocytosis and Other Signaling Events 7
CRMP and Disease 8
Concluding Remarks 9

2. GTPASES IN SEMAPHORIN SIGNALING 12

Andreas W. Püschel

Summary 12
Introduction 12
The Function of GTPases for Semaphorin Signaling 13
Interaction of GTPases with Plexins 14
GTPases and Signaling by Plexin-A1 16
GTPases and Signaling by Plexin-B1 19
Invertebrate Semaphorins 20
Open Questions 20

3. INTRACELLULAR KINASES IN SEMAPHORIN SIGNALING 24

Aminul Ahmed and Britta J. Eickholt

Introduction 24
Semaphorins Regulate Actin Filament Dynamics by Controlling
  Cofilin Activity 25
Modulation of Semaphorin Responses by Neurotrophins 27
Regulation of PI3K Signaling by Semaphorins 28
Synergistic Control of CRMP-2 Phosphorylation by CDK-5
  and GSK3 Mediates Sema3A Function 29
Cdk5 and Sema3A-Mediated Increases in Axonal Transport 30

MAPK Signaling and the Control of Sema3A Induced Translation
of Axonal mRNA 31
Semaphorin Signaling Leading to Selective Cell Death Responses 32
Concluding Remarks 32

4. MICAL FLAVOPROTEIN MONOOXYGENASES: STRUCTURE, FUNCTION AND ROLE IN SEMAPHORIN SIGNALING 38

Sharon M. Kolk and R. Jeroen Pasterkamp

Introduction 38
The MICAL Family 39
MICALs in Semaphorin Signaling 41
A MICAL Connection to the Cytoskeleton? 43
MICALs: Redox Regulators of Axon Guidance Events? 46
Concluding Remarks 47

5. SIGNALING OF SECRETED SEMAPHORINS IN GROWTH CONE STEERING 52

Sangwoo Shim and Guo-li Ming

Introduction 52
In Vitro Neuronal Growth Cone Steering Assays 52
Receptor Complex in Mediating Growth Cone Turning Responses
to Class 3 Semaphorins 53
Intracellular Mediators for Class 3 Semaphorin-Induced Growth Cone Turning 55
Modulation of Growth Cone Turning Responses to Class 3 Semaphorins 57
Summary 57

6. MODULATION OF SEMAPHORIN SIGNALING BY IG SUPERFAMILY CELL ADHESION MOLECULES 61

Ahmad Bechara, Julien Falk, Frédéric Moret and Valérie Castellani

Summary 61
Introduction 61
Soluble Forms of IgSFCAMs Convert Repulsive Responses to Class III
Semaphorins into Attraction 62
Transmembrane Forms of IgCAMs Are Components of Class III
Semaphorin Receptors 64
Molecular Interactions Underlying the Modulation of Semaphorin Signaling
by Soluble IgSFCAMs 64
Receptor Internalization and Modulation of Sema3A Signaling 65
Biological Contexts for Regulation of Semaphorin Signals by IgSFCAMS 66
Conclusions 68

7. PROTEOGLYCANS AS MODULATORS OF AXON GUIDANCE CUE FUNCTION 73

Joris de Wit and Joost Verhaagen

Introduction 73
Role of Heparan Sulfate Proteoglycans in Axon Guidance 74
Role of Chondroitin Sulfate Proteoglycans in Axon Guidance 80
Proteoglycans and Guidance Molecules in the Regeneration of Adult Neurons 82
Concluding Remarks 84

8. SEMAPHORIN SIGNALS IN CELL ADHESION AND CELL MIGRATION: FUNCTIONAL ROLE AND MOLECULAR MECHANISMS 90

Andrea Casazza, Pietro Fazzari and Luca Tamagnone

Introduction 90
Functional Role of Semaphorins in Cell Adhesion and Cell Migration 91
Signaling Molecules Mediating Semaphorin Function in Cell Adhesion and Migration 98
Some Open Questions 102
Conclusions 103

9. SEMAPHORIN SIGNALING DURING CARDIAC DEVELOPMENT 109

Toshihiko Toyofuku and Hitoshi Kikutani

Cardiac Morphogenesis: An Overview 109
Sema6D-Plexin-A1 Axis in Cardiac Morphogenesis 111
Semaphorin Signaling in Vascular Connections to the Heart 112
Signals of Sema3A in Endothelial Cell Migration 115
Summary and Perspectives 115

10. SEMAPHORIN SIGNALING IN VASCULAR AND TUMOR BIOLOGY 118

Gera Neufeld, Tali Lange, Asya Varshavsky and Ofra Kessler

Receptors Belonging to the Neuropilin and Plexin Families and Their Semaphorin Ligands 118
The Role of the Neuropilins in VEGF Signaling 123
The Role of Class-3 Semaphorins in the Control of Angiogenesis and Tumor Progression 124
Cell Surface Attached Semaphorins as Modulators of Angiogenesis and Tumor Progression 125
Semaphorins as Direct Regulators of Tumor Cell Behavior 127
Conclusions 127

11. SEMAPHORIN SIGNALING IN THE IMMUNE SYSTEM 132
Vincent Potiron, Patrick Nasarre, Joëlle Roche, Cynthia Healy and Laurence Boumsell

Introduction 132
CD100/SEMA4D 133
SEMA4A 137
Viral Semaphorins and SEMA7A / CD108 139
Other Immune Semaphorins and Related Proteins 141
Semaphorins in Lymphoid Disorders 142
Conclusion 143
Abbreviations 144

INDEX 145

CHAPTER 1

The CRMP Family of Proteins and Their Role in Sema3A Signaling

Eric F. Schmidt and Stephen M. Strittmatter*

Abstract

The CRMP proteins were originally identified as mediators of Sema3A signaling and neuronal differentiation. Much has been learned about the mechanism by which CRMPs regulate cellular responses to Sema3A. In this review, the evidence for CRMP as a component of the Sema3A signaling cascade and the modulation of CRMP by plexin and phosphorylation are considered. In addition, current knowledge of the function of CRMP in a variety of cellular processes, including regulation of the cytoskeleton and endocytosis, is discussed in relationship to the mechanisms of axonal growth cone Sema3A response.

The secreted protein Sema3A (collapsin-1) was the first identified vertebrate semaphorin. Sema3A acts primarily as a repulsive axon guidance cue, and can cause a dramatic collapse of the growth cone lamellipodium. This process results from the redistribution of the F-actin cytoskeleton[1,2] and endocytosis of the growth cone cell membrane.[2-4] Neuropilin-1 (NP1) and members of the class A plexins (PlexA) form a Sema3A receptor complex, with NP1 serving as a high-affinity ligand binding partner, and PlexA transducing the signal into the cell via its large intracellular domain. Although the effect of Sema3A on growth cones was first described nearly 15 years ago, the intracellular signaling pathways that lead to the cellular effects have only recently begun to be understood. Monomeric G-proteins, various kinases, the redox protein, MICAL, and protein turnover have all been implicated in PlexA transduction. In addition, the collapsin-response-mediator protein (CRMP) family of cytosolic phosphoproteins plays a crucial role in Sema3A/NP1/PlexA signal transduction. Current knowledge regarding CRMP functions are reviewed here.

Properties and Expression of CRMPs

A number of CRMP genes were identified independently in different species around the same time, and were named according to their method of discovery. CRMPs are also known as turned on after division (TOAD-64),[5] dihydropyrimidinase related protein (DRP),[6] unc33 like protein (Ulip),[7] and TUC (TOAD64/Ulip/CRMP).[8] Five vertebrate CRMP genes (CRMP1-5) have been identified, while the *Drosophila* genome appears to encode for only a single CRMP. CRMP1-4 share ~75% protein sequence identity with each other, however CRMP5 (also referred to as CRAM) is only 50-51% homologous. CRMPs share a high sequence homology with the *C. elegans unc-33* gene,[5,7,9,10] although two other nematode genes, CeCRMP1 and 2, have been classified in the CRMP family.[11] In addition, mammalian CRMP1, 2, and 4 appear to undergo alternative splicing.[12,13] CRMP isoforms strongly interact with

---

*Corresponding Author: Stephen M. Strittmatter—Department of Neurology, Yale University School of Medicine, New Haven, Connecticut, U.S.A. Email: stephen.strittmatter@yale.edu

Semaphorins: Receptor and Intracellular Signaling Mechanisms, edited by R. Jeroen Pasterkamp.
©2007 Landes Bioscience and Springer Science+Business Media.

each other and exist as heterotetramers when purified from brain.[14] Specificity exists for the hetero-oligomerization in that different isoforms have varying affinities for each other.[14] Information obtained from the examination of the crystal structure of CRMP1 homotetramers reveals that this specificity is likely due to differential polar and hydrophobic residues between isoforms at the two oligomerization interfaces.[15] CRMP1-4 genes share a high sequence homology (60%) with the liver dihydropyrimidinase (DHPase) and structural similarity with members of the metal-dependent amidohydrolases, both of which form stable tetramers. However, none of the CRMP isoforms demonstrate any enzymatic activity, likely due to the fact that they lack crucial His residues which coordinate binding of a metal atom at the active site of amidohydrolase enzymes.[6,14,15]

CRMPs were discovered to be one of the first proteins expressed in newly born neurons in the developing brain,[5] and CRMP2 expression has been shown to be induced by factors that promote neuronal differentiation such as noggin, chordin, GDNF, and FGF.[16-18] Not surprisingly, CRMPs are most highly expressed during the neurogenic period of brain development and expression peaks during the period of axon growth.[19] In addition, CRMP1, 2 and 5 are expressed in immature interneurons in the adult olfactory bulb,[20] a site of ongoing neurogenesis in adulthood.[21] The expression of CRMPs is restricted primarily to the nervous system, however some isoforms show a differential pattern of expression in various nervous system structures.[19,22] CRMP2, and to some extent CRMP3, are expressed in mature neurons at low levels. These expression patterns, when taken together with the fact that CRMPs form heterotetramers, imply that oligomers consisting of different combinations of monomeric isoforms may have different functional effects in various cell types.

The significant sequence similarity of CRMPs with the worm *unc-33* gene implies a role for CRMP in axon growth and morphology, since *unc-33* mutants display severe axonal abnormalities.[23,24] Also, overexpression of CRMP2 induces ectopic axon formation in cultured hippocampal cells.[25] Although CRMP is a cytosolic protein, a significant fraction has been shown to be tightly associated with the cell membrane.[5,26] This membrane-associated pool of CRMP is enriched at the leading edge of the growth cone lamellipodium and filopodia, further supporting a role in axon outgrowth and guidance.[5] All CRMP isoforms continue to be expressed through the period of axon pathfinding and synaptogenesis,[19] and CRMP expression is induced in sprouting fibers after injury in both the central and peripheral nervous system.[5,27,28]

Evidence For CRMPs in Sema3A Signaling

The first evidence for the involvement of CRMPs in Sema3A signaling came from the identification of chick CRMP-62 (CRMP2) in an expression cloning screen looking for Sema3A signaling components.[10] Injection of CRMP-62 (CRMP2) RNA rendered *Xenopus* oocytes electrophysiologically responsive to Sema3A (collapsin-1). Further, treatment of DRG neurons with antibodies developed against an N-terminal region (a.a. 30-48) of CRMP-62 blocked the ability of Sema3A to collapse their growth cones.

Sema3A signaling can be reconstituted in a nonneuronal heterologous system.[29,30] COS7 cells overexpressing PlexA1 and NP1 contract when Sema3A is applied to the media. This effect is easily quantified using alkaline phosphatase (AP)-tagged ligand and measuring the surface area of the cells with bound AP. AP-Sema3F treatment fails to cause contraction of PlexA1/NP1-expressing cells, demonstrating the specificity of the assay. COS7 cells overexpressing CRMPs in addition to the PlexA1/NP1 receptor components undergo contraction at a much more rapid rate than cells in which the receptors are expressed alone.[15] Therefore, CRMP proteins are able to facilitate Sema3A-mediated morphological changes in nonneuronal cells, further suggesting that they play a role in the signaling pathway (Fig. 1).

In addition, CRMP1-4 are all able to form a complex with the Sema3A receptor PlexA1 in transfected nonneuronal cells.[15] Although the presence of NP1 attenuates this interaction, stimulation with Sema3A reestablishes the complex. This is a specific effect since treatment with Sema3F is unable to promote PlexA1-CRMP interactions in the presence of NP1.

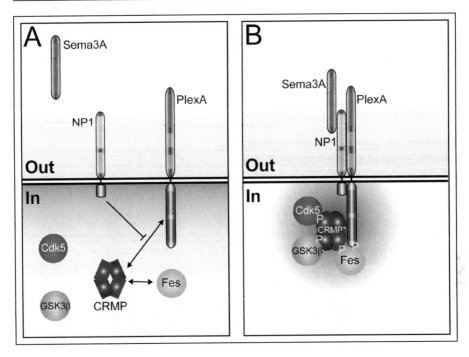

Figure 1. Schematic depicting the regulation of CRMP by Sema3A signaling. A) CRMP is a cytosolic phosphoprotein that exists as a heterotetramer and can bind to the kinase Fes. In the absence of Sema3A, NP1 blocks CRMP-PlexA interactions. B) Upon Sema3A stimulation, the CRMP tetramer is recruited to the cytoplasmic domain of PlexA, undergoes a conformational change (oval shape), and is phosphorylated by Cdk5, GSK3β, and Fes. These events lead to a change in the CRMP activation state subsequently mediating cellular responses to Sema3A.

Together with the fact that CRMPs facilitate COS7 contraction, these data demonstrate that PlexA1 and CRMP are able to initiate a Sema3A signaling cascade in nonneuronal cells.

What happens to CRMP upon Plexin binding? CRMPs may undergo a conformational change that leads to CRMP activation. Evidence for this is found in the observation that mutation of residues at the N-terminal region of the murine CRMP1 (a.a. 49-56) causes the protein to become constitutively active.[15] Overexpression of this mutant CRMP1 leads to the contraction of COS7 cells in the absence of PlexA1 and/or Sema3A stimulation, and the reduction of neurite outgrowth and Sema3A responsiveness in DRG neurons and explants. The expression of this mutant protein therefore mimics the cellular responses to Sema3A. Mutation of nearby residues (a.a. 38, 39, 41, 43) causes partial activation such that Plexin, but not Sema3A, is required for cell contraction (Fig. 2). Consistent with the mutation studies, antibody binding to residues 30-48 blocks Sema3A growth cone collapse,[10] suggesting that the N-terminal region between residues 30-56 plays an important role in mediating the activity of CRMPs in Sema3A signaling.

Phosphorylation of CRMP proteins at the C-terminal region may also be important in Sema3A signaling. A number of studies have shown the activity of a variety of kinases, including fyn, cdk5, GSK3β, LIMK, and Fes are important mediators of cellular responses to Sema3A[31-34] (Fig. 1). PlexA1 and A2 are constitutively bound to the src family tyrosine kinase, fyn, in a kinase-independent manner. Stimulation with Sema3A causes fyn activation and leads to the recruitment of the serine/threonine kinase, Cdk5, into the complex and activation of its kinase activity.[34] The blockade of Cdk5 kinase activity attenuates DRG growth cone collapse

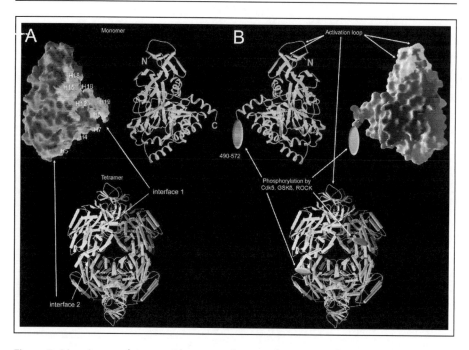

Figure 2. Mapping regulatory residues onto the crystal structure of CRMP1. A) The crystal structure of a CRMP1 monomer (top) and tetramer (bottom). The surface residues contributing to tetramerization are highlighted in orange (interface 1) and blue (interface 2) on a GRASP diagram of the CRMP1 monomer (top left). A ribbon diagram depicting α-helices and β-strands are shown for the CRMP1 monomer (top right). In the tetramer structure individual monomers are colored in blue, pink, cyan, and yellow with oligomerization interfaces 1 and 2 indicated. B) Residues involved in the regulation of CRMP activity are mapped onto the CRMP1 structure. (B, top left) Ribbon diagram of CRMP1 monomer is reflected 180° from that in A, and the C-terminal 80 amino acids are represented by the oval. The N-terminal "activation loop" is highlighted in pink on the GRASP diagram (upper right). All of the phosphorylation sites of CRMP are located on the C-terminal 80 amino acids that are not included in the crystal structure. A color version of this figure is available online at www.Eurekah.com.

in response to Sema3A.[34-36] CRMP1, 2, 4, and 5 are all phosphorylated by Cdk5 at Ser522,[36] and this can occur in response to Sema3A in vitro and in vivo.[35,36] The growth cones of neurons overexpressing a mutant CRMP2, in which Ser522 is replaced by alanine (S522A), fail to collapse when Sema3A is applied.[35,36]

The phosphorylation of CRMPs at Ser522 allows for the subsequent phosphorylation of CRMP2 and CRMP4 at Ser518, Thr509, and Thr514 by the serine/threonine kinase GSK3β.[35-38] It was previously shown that Sema3A activates GSK3β at the leading edge of growth cones, and this activation is required for proper responses to SemaA.[32] Overexpression of CRMP2 mutants T509A, T514A, S518A in neurons resulted in moderate decreases in Sema3A-induced growth cone collapse,[35,36] suggesting that single mutations are not sufficient to completely block Sema3A signaling. This may be due to the fact that the other phosphorylation sites are intact, or that CRMPs are not the primary targets of Sema3A-dependent GSK3β signaling. However, the regulation of CRMP by GSK3β phosphorylation may play an important role in Alzheimer's disease (AD), since CRMP2 is hyper-phosphorylated at Thr509, Ser518, and Ser522 in neurofibrilary tangles[39] and in amyloid precursor protein intracellular domain (AICD) transgenic mice.[40]

The nonreceptor tyrosine kinase, Fes, was copurified with CRMP5 (CRAM) from mouse brain.[33] Fes was able to phosphorylate all 5 CRMP proteins and PlexA1 in a constitutive manner, moreover, this effect was enhanced by Sema3A. The coexpression of Fes and PlexA1 in COS7 cells led to cell contraction, suggesting that Fes may be able to activate PlexA1. NP1 was able to block Plexin-Fes interactions, similar to the observations with CRMP-Plexin interactions, and this effect was overcome by Sema3A.[33] Further work needs to be done to identify the tyrosine residues on CRMP and Plexin that are putatively targeted by Fes, and to determine the functional consequences of Fes kinase activity.

The function of CRMP in axon guidance may not be restricted to Sema3A signaling. The growth cone collapsing factor LPA as well as the inhibitory guidance cue ephrin-A5 are both able to stimulate phosphorylation of CRMP2 at Thr555 by Rho kinase (ROCK), however Sema3A fails to do so.[41,42] Rho kinase is known to regulate neurite outgrowth and growth cone motility downstream of the GTPase, RhoA.[43] The kinase activity is specific for CRMP2 since Thr555 is not conserved in CRMP1,3,4, or 5.[41] Inhibitors to Rho kinase and expression of a kinase-dead mutant attenuate LPA mediated growth cone collapse, but have little to no effect on Sema3A growth cone collapse.[41] Further, the nonphosphorylated CRMP2 mutant T555A protects growth cones from the collapsing activity of both LPA and ephrinA5.[41,42] These observations demonstrate that CRMPs may be differentially regulated by multiple axon guidance cues, and thus be an important mediator of cellular responses to a variety of signals.

Examination of the crystal structure of CRMP1 reveals two separate regulatory regions of the protein (Fig. 2). The residues involved in constitutive activity and those targeted by the function blocking antibody all map the N-terminal "upper lobe" of the CRMP1 monomer, termed the "activation loop" (Fig. 2B, top; ref. 15). CRMP tetramers are assembled such that the activation loops are situated on the outer surface of the complex, allowing for regulation of other factors in the cytosol (Fig. 2B, bottom). The sites of phosphorylation by Cdk5, GSK3β, and ROCK are all localized to the C-terminus of the protein. Although the crystal structure does not include the C-terminal 80 amino acids, it seems likely that these residues also are located at the surface of the tetramer, supporting the hypothesis that phosphorylation may regulate tetramerization or that phosphorylation may be a reflection of CRMP oligomeric state (Fig. 2B, bottom).

The Regulation of the Cytoskeleton by CRMPs

Although it is clear that CRMP proteins are important for normal Sema3A responses in developing neurons, it is still not known what the downstream effectors of CRMP signaling may be. Rearrangement of the cytoskeleton is a well-characterized phenomenon resulting from Sema3A treatment.[1,2] Therefore one possibility is that CRMP2 may link Sema3A receptors to the cytoskeleton either directly or via factors known to regulate cytoskeletal dynamics (Fig. 3).

A growing body of literature suggests that CRMP2 plays a role in neuronal polarity by regulating microtubule dynamics.[38,44-46] CRMP2 is highly enriched in the growing axons of dissociated hippocampal cells, and when overexpressed, induces the formation of multiple axon-like processes.[25] Truncation of 24-191 amino acids at the C-terminus of CRMP2 prevents this axogenesis. CRMP2 colocalizes with microtubules in neuronal cells lines and fibroblasts, and CRMP1-4 can all bind to tubulin in vitro and in brain.[44-46] Furthermore, CRMP2 can induce microtubule assembly by binding to α- and β-tubulin heterodimers, an effect mediated by residues 323-381.[44] CRMP2's effect on axon formation is dependent upon the ability to promote microtubule assembly, since the overexpression of a mutant CRMP (Δ323-381) and the disruption of CRMP-microtubule interactions both fail to promote axon specification.[44]

A possible mechanism of Sema3A signaling may be an alteration of microtubule dynamics mediated by CRMP. Phosphorylation of Thr514 by GSK3β abolishes the ability of CRMP2 to bind to tubulin heterodimers.[38] Using a pThr514-specific antibody, the authors show that unphosphorylated CRMP2 is enriched at the leading edge of the growth cone. Expression of a

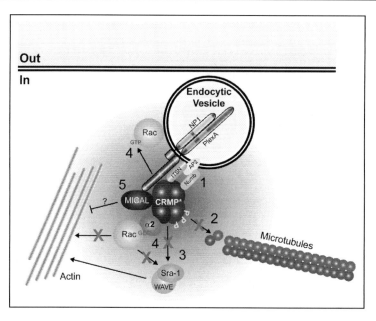

Figure 3. A model for CRMP-mediated cellular responses to Sema3A. 1) CRMP binds to the AP2 vesicle adaptor complex proteins intersectin (ITSN) and Numb to facilitate endocytosis of the growth cone membrane and PlexA/NP1 Sema3A receptor complex. 2) Phosphorylation of CRMP by GSK3β blocks the ability of CRMP to bind to tubulin dimers and subsequently prevents microtubule polymerization. 3) The Sra-1/WAVE complex, effectors of Rac1, regulate actin dynamics and are recruited to the growing axon by their interaction with CRMP. Sema3A may prevent the interaction of CRMP with the Sra-1/WAVE complex, thus attenuating recruitment and polymerization of actin in the growth cone. 4) CRMP interacts with the Rac1 GAP α2-chimaerin. The GAP activity may then lead to the inactivation of a pool of Rac1 mediating actin dynamics and an inhibition of the Sra-1/WAVE complex. Another pool of Rac1 is activated by PlexA and promotes endocytosis (top). 5) CRMP binds to the Plexin-interacting protein, MICAL, and may modulate its oxidoreductase enzymatic activity. A red "X" indicates interactions or processes that may be blocked by Sema3A signaling. Refer to text for details. A color version of this figure is available online at www.Eurekah.com.

nonphosphorylated CRMP2 mutant (T514A) causes an increase in axon length and branching relative to untreated or WT CRMP2 expression. The consitutively phosphorylated mutant T514D behaves similarly to WT CRMP2.[38] As discussed earlier, the phosphorylation of CRMPs by GSK3β is dependent on prior phosphorylation of Ser522 by Cdk5. Not surprisingly, the nonphosphorylated CRMP2 mutant S522A also shows enhanced association with microtubules and an increased ability to induce multiple axons.[36] Since the overexpression of S522A and T514A CRMP2 mutants leads to a reduction in Sema3A-induced growth cone collapse,[35,36] a model linking Sema3A to microtubule dynamics via CRMPs is supported. Interestingly, the phosphorylation of CRMP2 by Rho kinase also decreases the ability of CRMP2 to bind to tubulin dimers, supporting the idea that other guidance cues may converge on this pathway.[42]

Although microtubules regulate growth cone steering in response to axon guidance cues,[47] their polymerization state remains relatively unchanged immediately following Sema3A treatment.[1] It is possible that the CRMP2-dependent regulation of microtubules mediates longer term responses to Sema3A, such as growth cone turning.

During Sema3A-mediated growth cone collapse, a dramatic rearrangement of the actin cytoskeleton occurs.[1,2] A few studies have shown that a pool of CRMPs colocalize with actin in growth cones and neuronal cell lines.[13,24,46,48] It is therefore possible that CRMP activation modulates changes in actin dynamics. It is unlikely that such an effect is mediated by a direct binding of CRMP to actin, since there is no evidence that such an interaction occurs. However, it is possible that CRMP may be linked to the actin cytoskeleton by other proteins.

Key regulators of the actin cytoskeleton are Rho family GTPases. G protein signaling has been shown to play a large role in the cellular response to semaphorins.[49] Rac1, Rnd1, and RhoD have all specifically been shown to mediate class A plexin signaling.[50-53] CRMP2 is able to modulate the activities of Rac1 and RhoA. Expression of constitutively active RhoA (RhoA V14) in fibroblasts generally leads to the formation of stress fibers and focal adhesions. However, when CRMP2 is coexpressed with RhoA V14, there was a loss of stress fibers and increased formation of ruffles and microspikes, features usually associated with Rac1 activity. Conversely, when CRMP2 was coexpressed with the constitutively active Rac1 mutant (Rac1 V12), the cells lacked ruffles and developed stress fibers and some focal adhesions.[54] A similar switch in morphology was observed in neuroblastoma cells. Therefore, CRMP2 may act as a toggle between Rho GTPases. However, this activity of CRMP2 was dependent on phosphorylation of Thr555 by Rho kinase, which does not occur during Sema3A signaling.[41,54] It is not clear whether the phosphorylation of CRMP2 and/or other CRMP family members by Cdk5 and GSK3β can promote this activity.

The Sra1 protein is a downstream effector of Rac1 signaling, linking activated Rac1 to actin polymerization.[55] CRMP2 was shown to recruit a complex consisting of Sra1 and WAVE, another mediator of actin dynamics, to growth cones of cultured hippocampal neurons to promote axonal differentiation.[56] It will be interesting to determine whether this process can be regulated by Sema3A since it provides a direct link between Plexins, CRMP, and Rac1 regulation of the cytoskeleton. Since Sra1/WAVE promotes axon outgrowth, a possibility is that Sema3A attenuates the affinity of CRMP2 to associate with this complex.

The activity of Rho GTPases is dependent upon their nucleotide-binding state such that they are active when bound to GTP and inactive when bound to GDP.[57] Three classes of proteins exist that regulate this status: guanine nucleotide exchange factors (GEFs), which facilitate the exchange of GDP for GTP, GTPase activating proteins (GAPs), which promote GDP binding, and guanine dissociation inhibitors (GDIs), which also promote the GDP-bound state.[57] Recently, it was found that Sema3A promotes the dissociation of PlexA1 from the Rac1 GEF FARP2, and thus activation of Rac1 signaling.[51] CRMP2 has been shown to interact with the SH2 domain of the Rac1 GAP α2-chimaerin.[35] The significance of this interaction on the regulation of the GAP activity remains unclear, although active α2-chimaerin promotes CRMP2 association and the overexpression of GAP-inactive mutants block Sema3A-mediated growth cone collapse. These observations contradict studies showing the requirement for active Rac1 for Sema3A collapse of growth cones.[50,51] However, an initial, transient inactivation of Rac1 was seen immediately following an ephrinA2 or Sema3A challenge, followed by an increase in activity.[4] It is possible that the GAP activity of α2-chimaerin may mediate the initial decrease of Rac function, or that separate pools of Rac are differentially regulated (see below).

CRMP, Endocytosis and Other Signaling Events

In addition to disassembly of the cytoskeleton, Sema3A and other growth cone collapsing factors promote endocytosis of the growth cone membrane.[2-4] Within 10 minutes after treatment with Sema3A, PlexA1 and NP1 are recruited to vesicles inside the growth cone that also label with Rac1 and F-actin.[2] This process is dependent on the cell adhesion molecule L1 and Rac1 activation.[3,4] It is possible that the pool of Rac1 mediating endocytosis is separate from that regulating actin dynamics, partially explaining the discrepancies of the requirement of both Rac1 GEF and GAP activities in Sema3A signaling.

The long form of CRMP4 (TUC4b) was observed to be enriched in SV2 immunoreactive vesicles in the body of the growth cone.[13] Further, CRMP proteins were found to bind to the SH3A and SH3E domains of intersectin. Intersectin is a GEF for cdc42 and has been shown to promote N-WASP dependent actin polymerization.[58] However, it also associates with the AP2 adaptor complex of clathrin-coated vesicles in neurons[59] and may mediate vesicle dynamics.

Therefore, CRMP appears to link Plexin signaling to the endocytotic pathway. This hypothesis is further supported by the observation that CRMP2 also interacts with Numb and α-adaptin, both of which are a part of the AP2 adaptor complex and involved in clathrin-dependent endocytosis.[60,61] The binding of CRMP2 to Numb was required for normal endocytosis of L1 in growing hippocampal neurites. Overexpression of a CRMP2 mutant unable to bind to Numb, or knocking down expression of CRMP2 by RNAi was sufficient to attenuate L1 endocytosis.[61] The intersectin and Numb studies demonstrate that CRMP proteins provide a direct link between the intracellular domain of PlexA1 and the endocytotic machinery.

A number of other proteins have been shown to interact with CRMP, although the implication of these interactions in the context of Sema3A signaling needs to be elucidated. It was recently shown that CRMP2 negatively regulates phospholipase D2 (PLD2) activity, and Sema3A is able to inhibit PLD2 activity in PC12 cells.[62] Although the functional significance of this in growth cones is elusive, a number of intriguing possibilities exist since PLD2 is localized at the distal tips of neurons and has been implicated in regulating actin dynamics and receptor-mediated endocytosis.[62-64]

The flavin monooxygenase molecule interacting with CasL (MICAL), binds directly to *Drosophila* PlexA and is required for Sema1a signaling.[65] The proper axon guidance of fly motor neurons and Sema3A-mediated repulsion of mammalian DRG neurons is dependent on enzymatic activity.[65,66] The C-terminal half of MICAL has a number of conserved protein-protein interaction domains and signaling motifs, suggesting a complex role in cell signaling. Although no direct substrates for MICAL have been identified, it has been speculated that cytoskeletal proteins are potential targets.[65,67] Our unpublished observations indicate that CRMP1-4 can to bind to mammalian MICAL1. Whether CRMP is a substrate for MICAL or regulates MICAL activity is not yet clear. An intriguing possibility is that CRMP binds to and presents substrates to MICAL, since CRMP tetramers structurally resemble dihydropyrimidinase but lack enzyme activity. Regardless, the possibility for CRMP-MICAL interactions suggests a large signaling complex may exist to mediate Plexin signaling.

CRMP and Disease

Understanding CRMP signaling has significant clinical implications. The formation of appropriate axonal pathways and synaptic connections during nervous system development is critical for normal function in adults. Disruption of Sema3A signaling or CRMP activity may lead to devastating developmental neurological disorders. In addition, it has been shown that SemaA and CRMP are upregulated and play a role in axon growth and regeneration in response to nervous system injury.[5,27,28]

CRMP proteins have also been implicated in a more common adult neurological disease. A highly phosphorylated form of CRMP2 was identified as a component of the neurofibrillary tangles associated with Alzheimer's disease.[39,68] Transgenic mice overexpressing amyloid precursor protein intracellular domain (AICD), the protein implicated in the pathogenesis of Alzheimer's disease (AD), demonstrate an elevated level of phosphorylated CRMP2, likely due to increased GSK3β activity.[40] It will be important to determine whether normal CRMP activity or disfunction of CRMP proteins play a role in AD pathogenesis. Interestingly, a processed form of Sema3A as well as PlexA1, PlexA2, and CRMP2 were copurified in a complex from the brains of AD patients, leading to the possibility that disregulation of semaphorin signaling in general may play a role in disease phenotypes.[69]

CRMPs are autoantigens for some paraneoplastic neurological disorders. Auto-antibodies against CRMP3 (anti-CV2) and CRMP5 (anti-CRMP5 IgG) have been found in patients with certain types of cancer.[70,71] Patients who are producing these antibodies commonly suffer from peripheral neuropathy, cerebellar degeneration, and optic neuritis, among other neurological pathologies.[72,73] It is unclear whether the neuropathies are a direct result of disruption of CRMP function by the antibodies or a secondary effect. Surprisingly, CRMP1 was shown to be expressed by breast cancer cells, although high levels of expression actually led to decreased motility, and therefore abrogated invasiveness.[74]

Concluding Remarks

The CRMP family of proteins plays a central role in nervous system development and pathology. It is clear that CRMP is regulated by Sema3A signaling, and this is likely to occur both by phosphorylation of certain residues and by conformational changes in CRMP structure. Evidence links CRMP to cytoskeletal dynamics, G-protein signaling, and endocytosis, all of which are essential steps leading to growth cone responses to semaphorins (Fig. 3). The relative degree to which CRMP contributes to each of these processes remains to be elucidated, however it is quite evident that CRMP is able to tether these events to semaphorin receptors. Finally, interactions of CRMP family members with a variety of other signaling molecules and the pathogenesis of CRMP regulation in disease imply a role for CRMP in numerous cellular processes in addition to axon outgrowth and guidance. Therefore, further understanding of CRMP function is likely to elucidate numerous aspects of physiology and pathology.

References

1. Fan J, Mansfield SG, Redmond T et al. The organization of F-actin and microtubules in growth cones exposed to a brain-derived collapsing factor. J Cell Biol 1993; 121:867-878.
2. Fournier AE, Nakamura F, Kawamoto S et al. Semaphorin3A enhances endocytosis at sites of receptor-F-actin colocalization during growth cone collapse. J Cell Biol 2000; 149:411-422.
3. Castellani V, Falk J, Rougon G. Semaphorin3A-induced receptor endocytosis during axon guidance responses is mediated by L1 CAM. Mol Cell Neurosci 2004; 26:89-100
4. Jurney WM, Gallo G, Letourneau PC et al. Rac1-mediated endocytosis during ephrin-A2- and semaphorin 3A-induced growth cone collapse. J Neurosci 2002; 22:6019-6028.
5. Minturn JE, Fryer HJ, Geschwind DH et al. TOAD-64, a gene expressed early in neuronal differentiation in the rat, is related to unc-33, a C. elegans gene involved in axon outgrowth. J Neurosci 1995; 15:6757-6766.
6. Hamajima N, Matsuda K, Sakata S et al. A novel gene family defined by human dihydropyrimidinase and three related proteins with differential tissue distribution. Gene 1996; 180:157-163.
7. Byk T, Dobransky T, Cifuentes-Diaz C et al. Identification and molecular characterization of Unc-33-like phosphoprotein (Ulip), a putative mammalian homolog of the axonal guidance-associated unc-33 gene product. J Neurosci 1996; 16:688-701.
8. Quinn CC, Gray GE, Hockfield S. A family of proteins implicated in axon guidance and outgrowth. J Neurobiol 1999; 41:158-164.
9. Gaetano C, Matsuo T, Thiele CJ. Identification and characterization of a retinoic acid-regulated human homologue of the unc-33-like phosphoprotein gene (hUlip) from neuroblastoma cells. J Biol Chem 1997; 272:12195-12201.
10. Goshima Y, Nakamura F, Strittmatter P et al. Collapsin-induced growth cone collapse mediated by an intracellular protein related to UNC-33. Nature 1995; 376:509-514.
11. Takemoto T, Sasaki Y, Hamajima N et al. Cloning and characterization of the Caenorhabditis elegans CeCRMP/DHP-1 and -2; common ancestors of CRMP and dihydropyrimidinase? Gene 2000; 261:259-267.
12. Leung T, Ng Y, Cheong A et al. p80 ROKalpha binding protein is a novel splice variant of CRMP-1 which associates with CRMP-2 and modulates RhoA-induced neuronal morphology. FEBS Lett 2002; 532:445-449.
13. Quinn CC, Chen E, Kinjo TG et al. TUC-4b, a novel TUC family variant, regulates neurite outgrowth and associates with vesicles in the growth cone. J Neurosci 2003; 23:2815-2823.
14. Wang LH, Strittmatter SM. Brain CRMP forms heterotetramers similar to liver dihydropyrimidinase. J Neurochem 1997; 69:2261-2269.
15. Deo RC, Schmidt EF, Elhabazi A et al. Structural bases for CRMP function in plexin-dependent semaphorin3A signaling. Embo J 2004; 23:9-22.
16. Kamata T, Daar IO, Subleski M et al. Xenopus CRMP-2 is an early response gene to neural induction. Brain Res Mol Brain Res 1998; 57:201-210.
17. Kodama Y, Murakumo Y, Ichihara M et al. Induction of CRMP-2 by GDNF and analysis of the CRMP-2 promoter region. Biochem Biophys Res Commun 2004; 320:108-115.
18. Tateossian H, Powles N, Dickinson R et al. Determination of downstream targets of FGF signaling using gene trap and cDNA subtractive approaches. Exp Cell Res 2004; 292:101-114.
19. Wang LH, Strittmatter SM. A family of rat CRMP genes is differentially expressed in the nervous system. J Neurosci 1996; 16:6197-6207.

20. Veyrac A, Giannetti N, Charrier E et al. Expression of collapsin response mediator proteins 1, 2 and 5 is differentially regulated in newly generated and mature neurons of the adult olfactory system. Eur J Neurosci 2005; 21:2635-2648.
21. Doetsch F, Hen R. Young and excitable: The function of new neurons in the adult mammalian brain. Curr Opin Neurobiol 2005; 15:121-128.
22. Fukada M, Watakabe I, Yuasa-Kawada J et al. Molecular characterization of CRMP5, a novel member of the collapsin response mediator protein family. J Biol Chem 2000; 275:37957-37965.
23. Li W, Herman RK, Shaw JE. Analysis of the Caenorhabditis elegans axonal guidance and outgrowth gene unc-33. Genetics 1992; 132:675-689.
24. Hotta A, Inatome R, Yuasa-Kawada J et al. Critical role of collapsin response mediator protein-associated molecule CRAM for filopodia and growth cone development in neurons. Mol Biol Cell 2005; 16:32-39.
25. Inagaki N, Chihara K, Arimura N et al. CRMP-2 induces axons in cultured hippocampal neurons. Nat Neurosci 2001; 4:781-782.
26. Rosslenbroich V, Dai L, Franken S et al. Subcellular localization of collapsin response mediator proteins to lipid rafts. Biochem Biophys Res Commun 2003; 305:392-399.
27. Pasterkamp RJ, De Winter F, Holtmaat AJ et al. Evidence for a role of the chemorepellent semaphorin III and its receptor neuropilin-1 in the regeneration of primary olfactory axons. J Neurosci 1998; 18:9962-9976.
28. Suzuki Y, Nakagomi S, Namikawa K et al. Collapsin response mediator protein-2 accelerates axon regeneration of nerve-injured motor neurons of rat. J Neurochem 2003; 86:1042-1050.
29. Takahashi T, Fournier A, Nakamura F et al. Plexin-neuropilin-1 complexes form functional semaphorin-3A receptors. Cell 1999; 99:59-69.
30. Suto F, Ito K, Uemura M et al. Plexin-a4 mediates axon-repulsive activities of both secreted and transmembrane semaphorins and plays roles in nerve fiber guidance. J Neurosci 2005; 25:3628-3637.
31. Aizawa H, Wakatsuki S, Ishii A et al. Phosphorylation of cofilin by LIM-kinase is necessary for semaphorin 3A-induced growth cone collapse. Nat Neurosci 2001; 4:367-373.
32. Eickholt BJ, Walsh FS, Doherty P. An inactive pool of GSK-3 at the leading edge of growth cones is implicated in Semaphorin 3A signaling. J Cell Biol 2002; 157:211-217.
33. Mitsui N, Inatome R, Takahashi S et al. Involvement of Fes/Fps tyrosine kinase in semaphorin3A signaling. Embo J 2002; 21:3274-3285.
34. Sasaki Y, Cheng C, Uchida Y et al. Fyn and Cdk5 mediate semaphorin-3A signaling, which is involved in regulation of dendrite orientation in cerebral cortex. Neuron 2002; 35:907-920.
35. Brown M, Jacobs T, Eickholt B et al. Alpha2-chimaerin, cyclin-dependent Kinase 5/p35, and its target collapsin response mediator protein-2 are essential components in semaphorin 3A-induced growth-cone collapse. J Neurosci 2004; 24:8994-9004.
36. Uchida Y, Ohshima T, Sasaki Y et al. Semaphorin3A signaling is mediated via sequential Cdk5 and GSK3beta phosphorylation of CRMP2: Implication of common phosphorylating mechanism underlying axon guidance and Alzheimer's disease. Genes Cells 2005; 10:165-179.
37. Cole AR, Knebel A, Morrice NA et al. GSK-3 phosphorylation of the Alzheimer epitope within collapsin response mediator proteins regulates axon elongation in primary neurons. J Biol Chem 2004; 279:50176-50180.
38. Yoshimura T, Kawano Y, Arimura N et al. GSK-3beta regulates phosphorylation of CRMP-2 and neuronal polarity. Cell 2005; 120:137-149.
39. Gu Y, Hamajima N, Ihara Y. Neurofibrillary tangle-associated collapsin response mediator protein-2 (CRMP-2) is highly phosphorylated on Thr-509, Ser-518, and Ser-522. Biochemistry 2000; 39:4267-4275.
40. Ryan KA, Pimplikar SW. Activation of GSK-3 and phosphorylation of CRMP2 in transgenic mice expressing APP intracellular domain. J Cell Biol 2005; 171:327-335.
41. Arimura N, Inagaki N, Chihara K et al. Phosphorylation of collapsin response mediator protein-2 by Rho-kinase. Evidence for two separate signaling pathways for growth cone collapse. J Biol Chem 2000; 275:23973-23980.
42. Arimura N, Menager C, Kawano Y et al. Phosphorylation by Rho kinase regulates CRMP-2 activity in growth cones. Mol Cell Biol 2005; 25:9973-9984.
43. Amano M, Fukata Y, Kaibuchi K. Regulation and functions of Rho-associated kinase. Exp Cell Res 2000; 261:44-51.
44. Fukata Y, Itoh TJ, Kimura T et al. CRMP-2 binds to tubulin heterodimers to promote microtubule assembly. Nat Cell Biol 2002; 4:583-591.
45. Gu Y, Ihara Y. Evidence that collapsin response mediator protein-2 is involved in the dynamics of microtubules. J Biol Chem 2000; 275:17917-17920.

46. Yuasa-Kawada J, Suzuki R, Kano F et al. Axonal morphogenesis controlled by antagonistic roles of two CRMP subtypes in microtubule organization. Eur J Neurosci 2003; 17:2329-2343.
47. Gordon-Weeks PR. Microtubules and growth cone function. J Neurobiol 2004; 58:70-83.
48. Rosslenbroich V, Dai L, Baader SL et al. Collapsin response mediator protein-4 regulates F-actin bundling. Exp Cell Res 2005; 310:434-444.
49. Liu BP, Strittmatter SM. Semaphorin-mediated axonal guidance via Rho-related G proteins. Curr Opin Cell Biol 2001; 13:619-626.
50. Jin Z, Strittmatter SM. Rac1 mediates collapsin-1-induced growth cone collapse. J Neurosci 1997; 17:6256-6263.
51. Toyofuku T, Yoshida J, Sugimoto T et al. FARP2 triggers signals for Sema3A-mediated axonal repulsion. Nat Neurosci 2005; 8:1712-1719.
52. Turner LJ, Nicholls S, Hall A. The activity of the plexin-A1 receptor is regulated by Rac. J Biol Chem 2004; 279:33199-33205.
53. Zanata SM, Hovatta I, Rohm B et al. Antagonistic effects of Rnd1 and RhoD GTPases regulate receptor activity in Semaphorin 3A-induced cytoskeletal collapse. J Neurosci 2002; 22:471-477.
54. Hall C, Brown M, Jacobs T et al. Collapsin response mediator protein switches RhoA and Rac1 morphology in N1E-115 neuroblastoma cells and is regulated by Rho kinase. J Biol Chem 2001; 276:43482-43486.
55. Kobayashi K, Kuroda S, Fukata M et al. p140Sra-1 (specifically Rac1-associated protein) is a novel specific target for Rac1 small GTPase. J Biol Chem 1998; 273:291-295.
56. Kawano Y, Yoshimura T, Tsuboi D et al. CRMP-2 is involved in kinesin-1-dependent transport of the Sra-1/WAVE1 complex and axon formation. Mol Cell Biol 2005; 25:9920-9935.
57. Symons M, Settleman J. Rho family GTPases: More than simple switches. Trends Cell Biol 2000; 10:415-419.
58. Hussain NK, Jenna S, Glogauer M et al. Endocytic protein intersectin-l regulates actin assembly via Cdc42 and N-WASP. Nat Cell Biol 2001; 3:927-932.
59. Hussain NK, Yamabhai M, Ramjaun AR et al. Splice variants of intersectin are components of the endocytic machinery in neurons and nonneuronal cells. J Biol Chem 1999; 274:15671-15677.
60. Santolini E, Puri C, Salcini AE et al. Numb is an endocytic protein. J Cell Biol 2000; 151:1345-1352.
61. Nishimura T, Fukata Y, Kato K et al. CRMP-2 regulates polarized Numb-mediated endocytosis for axon growth. Nat Cell Biol 2003; 5:819-826.
62. Lee S, Kim JH, Lee CS et al. Collapsin response mediator protein-2 inhibits neuronal phospholipase D(2) activity by direct interaction. J Biol Chem 2002; 277:6542-6549.
63. McDermott M, Wakelam MJ, Morris AJ. Phospholipase D. Biochem Cell Biol 2004; 82:225-253.
64. Shen Y, Xu L, Foster DA. Role for phospholipase D in receptor-mediated endocytosis. Mol Cell Biol 2001; 21:595-602.
65. Terman JR, Mao T, Pasterkamp RJ et al. MICALs, a family of conserved flavoprotein oxidoreductases, function in plexin-mediated axonal repulsion. Cell 2002; 109:887-900.
66. Pasterkamp RJ, Dai HN, Terman JR et al. MICAL flavoprotein monooxygenases: Expression during neural development and following spinal cord injuries in the rat. Mol Cell Neurosci 2005.
67. Siebold C, Berrow N, Walter TS et al. High-resolution structure of the catalytic region of MICAL (molecule interacting with CasL), a multidomain flavoenzyme-signaling molecule. Proc Natl Acad Sci USA 2005; 102:16836-16841.
68. Yoshida H, Watanabe A, Ihara Y. Collapsin response mediator protein-2 is associated with neurofibrillary tangles in Alzheimer's disease. J Biol Chem 1998; 273:9761-9768.
69. Good PF, Alapat D, Hsu A et al. A role for semaphorin 3A signaling in the degeneration of hippocampal neurons during Alzheimer's disease. J Neurochem 2004; 91:716-736.
70. Honnorat J, Byk T, Kusters I et al. Ulip/CRMP proteins are recognized by autoantibodies in paraneoplastic neurological syndromes. Eur J Neurosci 1999; 11:4226-4232.
71. Yu Z, Kryzer TJ, Griesmann GE et al. CRMP-5 neuronal autoantibody: Marker of lung cancer and thymoma-related autoimmunity. Ann Neurol 2001; 49:146-154.
72. Cross SA, Salomao DR, Parisi JE et al. Paraneoplastic autoimmune optic neuritis with retinitis defined by CRMP-5-IgG. Ann Neurol 2003; 54:38-50.
73. Honnorat J, Antoine JC, Derrington E et al. Antibodies to a subpopulation of glial cells and a 66 kDa developmental protein in patients with paraneoplastic neurological syndromes. J Neurol Neurosurg Psychiatry 1996; 61:270-278.
74. Shih JY, Yang SC, Hong TM et al. Collapsin response mediator protein-1 and the invasion and metastasis of cancer cells. J Natl Cancer Inst 2001; 93:1392-1400.

Chapter 2

GTPases in Semaphorin Signaling

Andreas W. Püschel*

Summary

A hallmark of semaphorin receptors is their interaction with multiple GTPases. Plexins, the signal transducing component of semaphorin receptors, directly associate with several GTPases. In addition, they not only recruit guaninine nucleotide exchange factors (GEFs) and GTPase activating proteins (GAPs) but also are the only known integral membrane proteins that show a catalytic activity as GAPs for small GTPases. GTPases function upstream of semaphorin receptors and regulate the activity of plexins through an interaction with the cytoplasmic domain. The association of Plexin-A1 (Sema3A receptor) or Plexin-B1 (Sema4D receptor) with the GTPase Rnd1 and ligand-dependent receptor clustering are required for their activity as R-Ras GAPs. The GTPases R-Ras and Rho function downstream of plexins and are required for the repulsive effects of semaphorins. In this review, I will focus on the role of GTPases in signaling by two plexins that have been analyzed in most detail, Plexin-A1 and Plexin-B1.

Introduction

Semaphorins act as repellent signals that induce the collapse of growth cones or change the trajectories of axons away from the cells that produce these membrane-bound or secreted proteins.[1] The effects of semaphorins are usually mediated by binding to a receptor complex consisting of neuropilin-1 (Nrp-1) or Nrp-2 as the ligand binding subunit and an A-type plexin as the signal-transducing subunit (secreted class 3 semaphorins)[2-6] or by a direct interaction with a plexin (Sema3E: Plexin-D1; Sema4D: Plexin-B1; Sema5A: Plexin-B3; Sema6D: Plexin-A1; Sema7A: Plexin-C1).[6-9] At least three cellular processes are involved in the repulsive effect of semaphorins: depolymerization of actin filaments,[10,11] loss of integrin-mediated adhesion,[12,13] and stimulation of endocytosis.[14-16] Several kinases, including Cdk5, Src family kinases, Fes, mitogen-activated kinases, and GSK3β have been implicated in the response to semaphorins.[17-27] However, a hallmark of the semaphorin receptors is their direct and indirect interaction with multiple GTPases. Plexins, the signaling components of semaphorin receptors, not only directly bind GTPases and recruit GEFs and GAPs but also act as GAPs themselves. In this review, I will focus on the role of GTPases in signaling by Plexin-A1 and Plexin-B1. Although GTPases probably are important also for other semaphorin receptors, very little is known about these in terms of their interaction with GTPases.

*Andreas W. Püschel—Abteilung Molekularbiologie, Institut für Allgemeine Zoologie und Genetik, Westfälische Wilhelms-Universität, Schloßplatz 5, 48149 Münster, Germany. Email: apuschel@uni-muenster.de

Semaphorins: Receptor and Intracellular Signaling Mechanisms, edited by R. Jeroen Pasterkamp. ©2007 Landes Bioscience and Springer Science+Business Media.

The Function of GTPases for Semaphorin Signaling

In the absence of ligands, plexins assume an autoinhibited conformation by the interaction of the N-terminal semaphorin domain with the C-terminal half of the extracellular domain.[28] Binding of the ligand to a plexin, either directly (Sema4D) or indirectly via a coreceptor like Nrp-1 (Sema3A) results in a conformational change that relieves the autoinhibition and activates the receptor. Deletion of the semaphorin domain or the entire extracellular domain results in the constitutive activation of plexins.[28,29] There are two mutually not exclusive possibilities how a conformational change in the extracellular domain leads to the activation of downstream signaling. Analogous to the activation of integrins, a conformational change could be transmitted through the membrane to the cytoplasmic domain.[30] Ligand binding to integrin α/β dimers results in a separation of the cytoplasmic tails of α and β integrins and the recruitment of talin.[31-33] Alternatively, receptor activation could result in the formation of receptor oligomers.[29] Independent of how ligand binding is communicated to the cytoplasmic domain, GTPases are central components of the signaling pathways coupled to plexins that translate receptor activation into changes of cellular behavior.

There are over 150 small GTPases of the Ras superfamily encoded by the mammalian genome which perform essential functions in many different cellular processes.[34] The small GTPases can be divided into 5 subfamilies: the Ras, Rho, Arf, Rab, and Ran GTPases. They act as molecular switches and activate downstream effectors when in the GTP-bound form. Their intrinsic GTPase activity allows them to cycle between the active, GTP-bound and the inactive, GDP-bound state. GEFs catalyze the exchange of GDP for GTP and, thereby, activate GTPases. GTP-bound GTPases activate downstream signaling by interacting with specific motifs in various effector proteins. GAPs terminate signaling by GTPases by stimulating their intrinsic GTPase activity. The structure and dynamics of the cytoskeleton is regulated by the Rho GTPases. Rho GTPases can be subdived into several subgroups (Cdc42, Rho, Rac, Rnd1, and RhoD).[35] While most Rho GTPases function according to the general principles outlined above, Rnd1 is an unusual member of the Rho GTPases because it is considered to be constitutively active due to its low intrinsic GTPase activity.[36]

The central position that members of the Rho subfamily occupy in the regulation of the cytoskeleton make them prime candidates for mediating the repulsive and attractive effects of axon guidance molecules. Work by several labs has shown that increased activity of Rho results in a reduction of neurite growth or neurite retraction while active Rac has the opposite effect.[37] Rho and Rac regulate each other antagonistically. In fibroblasts, Rac downregulates Rho activity by stimulating the production of reactive oxygen species.[38] Oxygen radicals inhibit phosphateses by oxidation of their catalytic center which results in an increase in Rho GAP phosphorylation and activity. Both Sema3A and Sema4D induce the repulsion or collapse of growth cones from different types of neurons.[1] It was expected that Rho is required for this effect similar to its role in the growth cone collapse induced by LPA and ephrins.[39,40] However, an involvement of Rho has been shown so far only in the case of the Sema4D receptor Plexin-B1.[41-44] Unexpectedly, dominant-negative Rac1N17 blocked the collapse by Sema3A while Rho does not appear to be involved in Sema3A signaling.[45-47]

Two culture systems are mainly used to investigate the role of GTPases in semaphorin signaling. Incubation of sensory or hippocampal neurons with Sema3A or Sema4D results in the rapid collapse of axonal growth cones within less than one hour.[10,48] A second assay is based on the reconstitution of functional receptors in COS 7 cells.[5] These do not express functional semaphorin receptors or respond to semaphorins but become responsive to Sema3A or Sema4D upon expression of Plexin-A1 and Neuropilin-1[5] or Plexin-B1,[43] respectively. This heterologous system mimics growth cone collapse and has been used to study the properties of semaphorin receptors. Using these assays, various pharmacological agents and dominant-negative constructs have been identified that interfere with the function of GTPases and block growth cone cell collapse. These in vitro systems together with genetic approaches in *Drosophila* and *C. elegans* provided strong evidence for the involvement of several GTPases in semaphorin signaling.

Interaction of GTPases with Plexins

The unique feature of semaphorin receptors is their interaction with multiple GTPases. In addition to R-Ras that serves as the substrate for the GAP activity of Plexin-A1 and -B1,[29,49,50] they also can directly interact with the Rho GTPases Rac (Plexin-A1 and -B1),[51-53] RhoD (Plexin-A1),[54] and Rnd1 (Plexin-A1 and -B1).[43] One of the first indications for a role of GTPases in semaphorin signaling was the observation that the cytoplasmic domain of plexins shows sequence similarity to GAPs specific for Ras GTPases.[52] This homology is divided into two highly conserved blocks (C1 and C2) that are separated by a sequence of variable size (V1) which is less well conserved between different plexins and essential for the binding of Rac1, Rnd1, and RhoD (Fig. 1). The minimal requirement for GTPase binding has not been determined precisely for Plexin-A1 and Plexin-B1. The shortest fragments of Plexin-A1 and -B1 that still bind to Rac1 comprise amino acids 1476 to 1623 (Plexin-A1)[55] and 1696 to 1919 (Plexin-B1),[53] respectively. A smaller fragment of Plexin-B1 (amino acid 1811-1910) is no longer able to bind Rac1.[53] Binding of Rac1, Rnd1, and RhoD is abolished by mutation of three

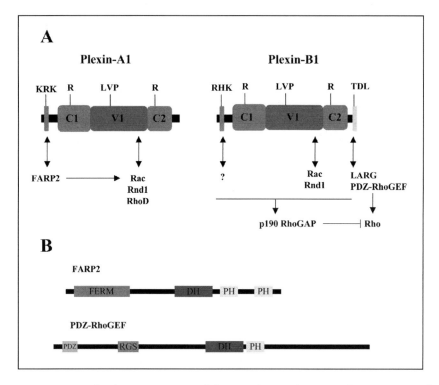

Figure 1. Direct and indirect interaction of the cytoplasmic domain with GTPases. A) The position of sequences relevant for the interaction with GTPases is shown for the cytoplasmic domains of Plexin-A1 (amino acids 1264 - 1894, accession no. NM_008881) and Plexin-B1 (amino acids 1495 - 2119, NM_172775). The conserved sequences showing homology to Ras GAPs (C1 and C2; Plexin-A1: amino acids 1379 - 1508 and 1727 - 1820; Plexin-B1: 1629 - 1759 and 1975) are separated by a less well conserved sequence V1 that is required for interaction with the indicated GTPases. C1 and C2 contain conserved arginine residues (R) that are essential for their function as GAPs. The juxtamembrane sequence KRK is required for the binding of FARP2 to Plexin-A1. A similar stretch of basic amino acids (RRK) is found also in Plexin-B1 but it not known if this motif interacts with a GEF. The C-terminus of Plexin-B1 contains the PDZ-binding motif TDL. B) The domain structure of the GEFs FAPR2 and PDZ-RhoGEF is shown.

residues in V1 (LVP; Plexin-A1: amino acid 1598-1600; Plexin-B1: 1849-1851).[53,54] It has been suggested that this sequence is part a CRIB-like (Cdc42/Rac interactive binding) motif.[53] However, the sequence similarity to CRIB motifs is very limited and mutation of a single proline residue that is highly conserved in CRIB sequences has no effect on GTPase binding.[54] Although mutation of the LVP motif in Plexin-A1 and -B1 abolishes the interaction with all Rho family GTPases, it still has to be shown directly that all GTPases bind to the same sequence.

There are contradictory reports in the literature concerning the interaction of Plexin-A1 and -B1 with Rac1 and Rnd1. While Zanata et al (2003) showed that Plexin-A1 binds Rnd1 but not Rac1,[54] others describe a strong interaction with Rac1 but could not detect one with Rnd1.[43,55] Oinuma et al (2003) detected a direct binding of Plexin-B1 to Rnd1, Rnd2, and weakly to Rnd3.[43] An interaction of Rnd1 with Plexin-B1 was also described by Rohm et al (2000).[52] However, this interaction was much weaker than that for Rac. While Rohm et al (2000) analyzed the interaction of plexins and Rnd1 in a GST pull-down assay using full length plexins expressed in 293T cells,[52] the other studies employed a filter binding assay and the isolated cytoplasmic domains.[43,55] More recently, a role for both Rac1 and Rnd1 in Plexin-A1 signaling was reported. Toyofuko et al (2005) showed that the interaction of Rnd1 and Plexin-A1 was observed only when Plexin-A1 was expressed alone but not after the coexpression with Nrp1.[50] This result confirms the existence of at least two conformations that differ in their ability to interact with GTPases. The described discrepancies may, therefore, reflect distinct conformations present under different experimental conditions.

Different experiments showed that Rho GTPases act upstream of the plexins and regulate their activity as receptors. Rnd1 and RhoD have antagonistic effects on the activity of Plexin-A1.[54] Coexpression of Rnd1 and Plexin-A1 in COS 7 cells triggers signaling by Plexin-A1 and results in a cell collapse in the absence of any ligand. By contrast, RhoD has the opposite effect and blocks Plexin-A1 activity both in COS 7 cells and in sensory neurons. Rac1 also regulates Plexin-A1 function by promoting the recruitment of Rnd1.[50] While Rnd1 has a similar function upstream of Plexin-A1 and Plexin-B1,[29,49] the role of Rac1 for Plexin-B1 signaling is less clear. Its effect on surface expression and receptor affinity[56] suggest that Rac1 also acts upstream of Plexin-B1 but may play a different role than that described for Plexin-A1. Similar to mammalian Plexin-B1, *Drosophila* Plexin-B binds active Rac. In addition, a different region of Plexin-B that is not conserved in mammalian plexins interacts with Rho.[57]

Plexins not only directly interact with several GTPases but also function as GAPs themselves.[29,49,50] The GAP homologies C1 and C2 include two arginine residues (Fig. 1) that correspond to the catalytic residues found in Ras GAPs and are conserved between all plexins and Ras GAPs.[52,58-60] The conserved arginine residues in C1 (R1429/1430 in Plexin-A1) correspond to the invariant arginine located in the finger-loop of Ras GAPs that inserts into the active center of the Ras GTPase.[52] The second conserved arginine residue (corresponding to R1746 in Plexin-A1) stabilizes the finger-loop in Ras GAPs. These arginine residues are essential for the activity of Ras GAPs and required for plexins to function as the signal-transducing subunit of semaphorin receptors.[29,49,50,52] A mutation of either amino acid suffices to completely block the ability of Plexin-A1 to induce the collapse of COS-7 cells.[52] Only recently, it was shown directly that the cytoplasmic domain of Plexin-A1 and -B1 interact with the GTPase R-Ras and act as a GAPs for it.[29,49,50]

The plexins are the first integral membrane proteins with an intracellular domain that has a catalytic activity as a GAP. The GAP activity is regulated by oligomerization and GTPases through changes in the conformation of the intracellular domain. The two GAP homology regions C1 and C2 interact with each other.[29,55] This interaction is regulated by the binding of GTPases to the insert region V1. Binding of Rac1 or Rnd1 to the V1 region of Plexin-A1 or -B1, respectively, disrupts the interaction between C1 and C2. In addition to Rnd1 recruitment, the clustering of plexins is required for GAP function.[29,49,50] The physiological relevance of this GAP activity was demonstrated by showing that R-Ras is required for the collapse induced by Sema3A and Sema4D through the Plexin-A/Neuropilin-1 and the Plexin-B1 receptors, respectively.[43,50]

GTPases and Signaling by Plexin-A1

Signaling by A-type plexins involves the sequential interaction of the three GTPases Rac1, Rnd1, and R-Ras with the cytoplasmic domain of A-type plexins.[29,49,50,54] In contrast to Plexin-B1 (see below), Rho does not appear to be required for Sema3A signaling.[55] Although the interaction with GTPases has been studied in detail only for Plexin-A1,[9] the high conservation of their cytoplasmic domain suggests that all A-type plexins function in a similar way. The interaction between GTPases and Plexin-A1 is modulated by the ligand-binding subunit Nrp-1. Nrp1 and Plexin-A1 constitutively associate to form a receptor complex[2-6] in which Plexin-A1 assumes a conformation that allows the recruitment of the Rac GEF FARP2 but not the binding of Rnd1.[9] While Rnd1 constitutively associates with the cytoplasmic domain of Plexin-A1 in the absence of Nrp-1, this interaction is completely dependent on Sema3A in the presence of Nrp-1. The GAP homology regions C1 and C2 (Fig. 1) interact with each other and probably assume a conformation without GAP activity.[55] The association requires the V1 region located between the conserved sequences. It remains to be investigated if C1 and C2 show an intra- or intermolecular interaction in the native receptor complex.

Rac1 and Rnd1 act upstream of Plexin-A1 (Fig. 2A). The first known intracellular event after ligand binding is the activation of Rac1 by FARP2.[50] A juxtamembrane sequence consisting of three basic amino acid residues (KRK) that is present in all A-type plexins is required for binding FARP2 through its FERM domain (Fig. 1). Binding of FARP2 to Plexin-A1 reduces its GEF activity. Efficient recruitment of FARP2 can be observed only in the presence of Nrp-1. Consistent with previous results,[61] the extracellular but not the cytoplasmic domain of Nrp-1 is required for FARP2 binding.[50] This suggests that Nrp-1 promotes a conformation of Plexin-A1 that allows the association of FARP2 but but not of Rnd1.

Within 10 minutes of binding to the receptor complex, Sema3A induces the dissociation of FARP2 from the Plexin-A1 and an increase in Rac activity (Fig. 2). The amount of GTP-bound Rac reaches maximal levels at 10-30 min after ligand binding and returns to basal levels at 60 min.[50] The activation of Rac is also observed with a Plexin-A1 mutant unable to bind Rnd1, showing that Rac1 acts upstream of Rnd1.[55] In the presence of Sema3A, FARP2 does not reassociate with Plexin-A1. Activation of Rac1 by FARP2 facilitates the association of Rnd1 with Plexin-A1 which is required for the GAP activity of Plexin-A1. The dependence of the association of Rnd1 with the Plexin-A1/Nrp-1 receptor complex on Rac1 can explain why dominant-negative Rac1N17 blocks the collapse of motor neuron growth cones and COS 7 cells by Sema3A.[46,55] RacN17 has no affect on the collapse mediated by a constitutively active Plexin-A1 mutant,[55] confirming that it acts upstream of Plexin-A1. Since Rac1 and Rnd1 bind to the same or overlapping sequences of Plexin-A1, it remains to be investigated how activation of Rac1 promotes Rnd1 binding.

Association with Rnd1 is required for Plexin-A1 to stimulate the GTPase activity for R-Ras but blocks basal GAP activity.[50] As orginally described for Plexin-B1, Plexins-A1 requires ligand-induced or artifical clustering (in vitro) in addition to the recruitment of Rnd1 to display its activity as a GAP. The resulting decrease in active R-Ras contributes to the suppression of integrin-mediated cell adhesion by Sema3A since R-Ras is a central regulator of integrin function.[62] After dissociation from Plexin-A1, FARP2 associates with and inhibits the type I γ phosphatidylinositol-4-phosphate 5-kinase (PIPKIγ661), enhancing the decrease in integrin-dependent adhesion.[50] PIPKIγ661 is targeted to focal adhesions through its association with talin and strengthens the binding of talin to β-integrin by PtdIns(4,5)P(2) production.[63,64] The GEF activity of FARP2 is suppressed when it forms a complex with PIPKIγ661, suggesting that this interaction also contributes to the inactivation of FAPR2 after disengagement from Plexin-A1.

The inhibition of Rnd1 binding to Plexin-A1 in the presence of Nrp-1 suggests the existence of two conformations that are interconvertable by interaction with active Rac1 (Fig. 3). The available data suggest that C1 and C2 associate to form an autoinhibited complex without GAP activity.[55] Active Rac1 can disrupt the interaction between C1 and C2 and presumably

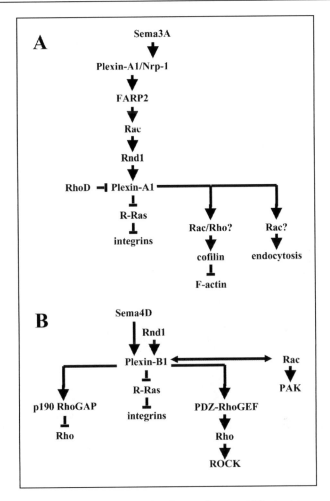

Figure 2. Signaling by plexins. A) Sema3A binding to the Nrp-1/Plexin-A1 receptor complex induces the dissociation and activation of FARP2. FARP2 activates Rac1 which facilitates the recruitment of Rnd1. Rnd1 binding is required for the GAP activity of clustered Plexin-A1. The GAP activity of Plexins-A1 reduces the amount of active R-Ras which results in a loss of integrin-dependent cell adhesion. Activation of other, less well characterized pathways results in the depolymerization of F-actin and the stimulation of endocytosis. B) Binding of Sema4D to Plexin-B1 stimulates GAP activity only when Rnd1 is associated with the cytoplasmic domain. The resulting inactivation of R-Ras leads to a loss of cell adhesion. PDZ-RhoGEF bound to the C-terminus of Plexin-B1 is activated and mediates an increase of GTP-bound Rho and ROCK activity. The competition of Plexin-B1 and PAK for binding of Rac may result in a decrease of PAK activity. Plexin-B1 also transiently associates with and activates p190 RhoGAP.

promotes a conformation that is able to bind Rnd1.[50,55] C1 and C2 each contain arginine residues that are essential for the function of Plexin-A1 and GAP activity.[50,52] If the molecular mechanism underlying the GAP function of plexins is similar to that of Ras GAPs, one has to assume that the arginine residue in C2 interacts with residues in C1 to form a functional catalytic center as described for Ras GAPs.[58-60] Thus, the sequential activity of Rac1 and Rnd1 may promote GAP activity by inducing a switch in the interactions between C1 and C2 (Fig.

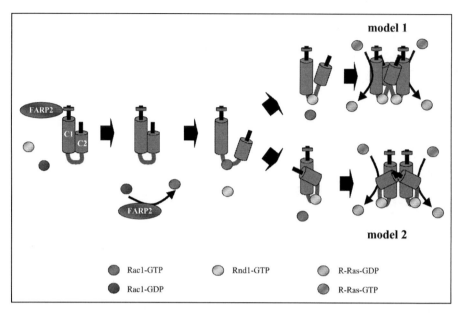

Figure 3. A model for the conformational changes of the cytoplasmic domain after receptor activation. The cytoplasmic domain of Plexin-A1 is shown schematically with the C1 and C2 regions (red), the GTPase-binding V1 region (blue), and the FAPR2 binding site (mangenta). In the unactivated state, the intra- or intermolecular association of C1 and C2 in the Nrp-1/Plexin-A1 complex does not allow the formation of an active R-Ras GAP and the recruitment of Rnd1. Activation of the receptor by Sema3A results in the dissociation of FARP2 and Rac activation. Interaction of Rac and Plexin-A1 disrupts the interaction of C1 and C2 and results in the recruitment of Rnd1 to the V1 region. Rnd1 may promote an additional conformational change and the inter- (model 1) or intramolecular (model 2) reassociation of C1 and C2 in a different orientation that displays GAP activity. The suggested reassociation of C1 and C2 remain to be confirmed experimentally. A color version of this figure is available online at http://www.Eurekah.com.

3). The requirement for an oligomerization of plexins suggests that, in the active GAP, the C1 and C2 regions of two different plexin molecules might interact (Fig. 3). This question, however, requires further analysis.

While the role of GTPases in regulating plexin function begins to be understood in some detail, little is known about the involvement of their effectors during growth cone collapse. A prominent effect of Sema3A during growth cone collapse is the depolymerization of F-actin. Sema3A induces the sequential phosphorylation and dephosphorylation of cofilin.[65] Cofilin is a key regulator of actin dynamics and phosphorylated by LIM kinase 1 (LIMK1). A dominant-negative LIMK1 blocks growth cone collapse by Sema3A. LIMK1 itself is regulated the GTPase effectors PAK and ROCK.[66] Dominant-negative Rac1 blocks Sema3A-induced growth cone collapse.[45,46] In addition, peptides derived from amino acids 17-32 of Rac1 inhibit Sema3-induced collapse as does a peptide from the CRIB-domain of PAK.[47] The stimulation of endocytosis during growth cone collapse by Sema3A requires Rac1.[15,16] However, it is possible that the block of endocytosis observed after Rac1 inhibition is a consequence of preventing the activation of plexins upstream of Rnd1. The results of Toyofuku et al (2005)[50] suggest that the effects of inhibiting Rac1 function may mainly reflect its role in promoting the recruitment of Rnd1. It remains to be investigated which function Rac may play in addition downstream of plexins, e.g., in regulating actin polymerization or endocytosis.

GTPases and Signaling by Plexin-B1

The extracellular domain of the integral membrane protein Sema4D binds directly to its receptor Plexin-B1. Sema4D induces the collapse of axonal growth cones from hippocampal neurons and the contraction of COS 7 cells coexpressing Plexin-B1 and Rnd1.[43,44] Four GTPases are involved in the response to Sema4D (Fig. 2B). Plexin-B1 directly interacts with the GTPases R-Ras, Rac1 and Rnd1. In addition, it regulates the activity of Rho through PDZ-RhoGEF/LARG[41,42,44,67] and p190 RhoGAP.[68]

Plexin B1 was the first plexin for which a GAP activity was demonstrated.[29,49] At least in a heterologous system, Rnd1 and Plexin-B1 associate constitutively.[43] Association with Rnd1 inhibits the basal GAP activity of Plexin-B1 both in vitro and after heterologous expression.[29,49] The formation of this complex is not affected by Sema4D. The association with Rnd1 is necessary but not sufficient for Plexin-B1 to stimulate the GTPase activity of R-Ras. Binding of Sema4D induces the rapid clustering of Plexin-B1 which peaks at 5 min after ligand addition.[29,49] In vitro, clustering of the isolated cytoplasmic domain by antibodies can mimic the effect of ligand binding. Both the binding of Rnd1 and clustering of Plexin-B1 are required for GAP activity. The stimulation of GAP activity by Sema4D transiently reduces the amount of active R-Ras. Since GTP-bound R-Ras activates integrins and stimulates integrin-based cell adhesion,[62] a reduction in integrin activity is an early event in the contraction of COS 7 cells induced by Sema4D and during the collapse of growth cones.

As described for Plexin-A1, the C1 and C2 domains interact with each other.[29] The association of C1 and C2 depends on V1, and is disrupted by Rnd1 binding to V1. It is presently not known if additional GTPases are required to facilitate the recruitment of Rnd1 as described for Plexin-A1. Plexin-B1 strongly interacts with Rac1 and this association is stimulated by Sema4D.[53] However, active Rac1 does not induce the contraction of COS 7 cells expressing Plexin-B1 or promote it after Sema4D addition.[43] Like Plexin-A1, the cytoplasmic domain of Plexin-B1 contains a juxtamembrane sequence of three basic amino acids residues but does not bind FARP2.[50] Thus, activtion of Rac may not be required or a different GEF assumes the role of FARP2 for Plexin-B1. Alternatively, Rac binding may link Plexin-B1 to PAK (Fig. 2B).[56] Plexin-B1 and PAK can compete for Rac binding, leading to a reduction in the Rac-dependent activation of PAK. Binding of Rac was suggested to sequester it and thereby decrease the interaction with its effector PAK. This competition for PAK was shown, however, only in a heterologous system. In addition, active Rac increases the amount of Plexin-B1 localized to the cell surface and enhances the binding of Sema4D to Plexin-B1 by a small increase in receptor affinity.[56] Consistent with these results, an intact Rac binding site is essential for the efficient expression of Plexin-B1 on the surface of transfected cells.

In addition to R-Ras, the regulation of Rho is important for Sema4D-induced collapse. The collapse of growth cones from hippocampal neurons by soluble Sema4D can be blocked by the Rho inhibitor C3 exoenzyme and the ROCK inhibitor Y-27632.[44] The contraction of COS 7 cells coexpressing Plexin-B1 and Rnd1 in response to Sema4D can also be blocked by dominant-negative RhoAN19 and Y-27632.[43] Sema4D binding to Plexin-B1 leads to an increase in Rho activity in a heterologous system when the receptor is coexpressed with the RhoGEFs PDZ-RhoGEF or LARG. Plexin-B1 constitutively associates with these RhoGEFs through the C-terminal PDZ-binding motiv that is present in all mammalian B-type plexins but absent from other plexins (Fig. 1).[44] This interaction is increased by Rnd1[43,44] but not stimulated by Sema4D.[41,42,67] A dominant-negative PDZ-RhoGEF construct blocks the Sema4D-induced contraction of COS 7 cells and the collapse of hippocampal growth cones but has no effect on the response to Sema3A. Expression of Rnd1 but not active Rac1 enhances the activation of RhoA by Sema4D in COS 7 cells expressing Plexin-B1 by increased binding of PDZ-RhoGEF. The interaction with PDZ-RhoGEF also promotes the translocation of Plexin-B1 to the plasma membrane.

Plexin-B1 transiently associates with p190 RhoGAP whose GAP activity is increased upon binding.[68] Barberis et al (2005) report that activation of Plexin-B1 by Sema4D is accompanied by a downregulation of Rho activity in some cell lines.[68] This observation contrasts with the increase in Rho-GTP levels which is mediated by PDZ-RhoGEF and LARG.[44] Differences in the cell lines and culture condition used may account for most of the differences between these studies. Another explanation put forward by Barberis et al (2005) is that Rho levels may be transiently decreased after receptor activation (at 5 - 30 min) but increased at later time points (60 min).[68] p190 RhoGAP is required for the reponse to semaphorins in epithelial and primary endothelial cells as well as in neuroblasts. However, the function of the decrease in Rho activity for the response to Sema4D remains to be elucidated.

At the moment, it cannot be excluded that Plexin-B1, like Plexin-A1, requires a coreceptor for binding or signaling since all experiments analyzing Sema4D receptors were done after expression of Plexin-B1 in cells and not with purified proteins.[6] In epithelial cells, Plexin-B1 associates with the hepatocyte growth factor receptor Met.[69] Sema4D stimulates the tyrosine kinase activity of Met which leads to the phosphorylation of Plexin-B1 and Met. Plexin-B1 also forms a complex with the receptor tyrosine kinase ErbB-2 after heterologous expression in HEK 293 cells and ErbB-2 is required for the activation of Rho.[70] Sema4D binding to the Plexin-B1/ErbB-2 complex stimulates kinase activity with 1 min of addition. Kinase activity peaks at 10 min and results in phosphorylation of both Plexin-B1 and PDZ-RhoGEF. Constitutively active Rac1 and Rnd1 enhance the phosphorylation of Plexin-B1 induced by Sema4D. Both Met and ErbB-2 are present in axonal growth cones of hippocampal neurons.[70] Dominant-negative mutants of Met and ErbB-2 block the collapse of growth cones by Sema4D but not by Sema3A. These results suggest that the kinases Met and ErbB-2 are involved in Sema4D signaling but their precise role in neurons remains to be investigated.

Invertebrate Semaphorins

The signaling by invertbrate plexins is not well understood. The high conservation of their cytoplasmic domains suggests that *Drosophila* and *C. elegans* plexins could also act as RasGAPs. It remains to be shown genetically or biochemically that the closest R-Ras homologs in flies (Ras64B, 69% identity) or worms (Ras-1, 65% identity) are indeed substrates for plexins. There are also no clear homologs for Rnd1 or RhoD in *Drosophila* and *C. elegans*. Other aspects of GTPase signaling will certainly differ. In genetic experiments, *Drosophila* Plexin-B suppresses Rac and enhances RhoA activity.[57] Plexin-B does not contain a recognizable PDZ-binding motif at its C-terminus but, unlike mammalian B-type plexins, interacts directly with Rho. Sequences essential for this interaction map to the V1 region that are not conserved.[57] As suggested for Plexin-B1, *Drosophila* Plexin-B may sequester Rac1 and thereby suppress PAK activity. Genetic interactions suggests that the A-type *C. elegans* Plexin-1 acts to sense the level of RHO (*rho-1*) and RAC GTPases (*mig-2* and *ced-10*) which determines the effect of semaphorins on the positioning of sensory ray 1.[71]

Open Questions

Although much progress has been made in dissecting the signaling cascades that mediate the effects of semaphorins and the role of GTPases in these pathways, we are still at the beginning to understand how activation of semaphorin receptors leads to growth cone collapse or changes in the trajectories of neurites. An important question that remains to be answered is the function of GTPases downstream of plexins and the identity of the effectors linked to them. Recent results revealed a centrol role for R-Ras in semaphorin signaling. Since R-Ras has been shown to be important for regulating integrin function, the inactivation of R-Ras by plexins could explain to a large extent the loss of cell adhesion during the growth cone collapse induced by semaphorins. It remains to be tested if R-Ras is the only target of plexins or if other GTPases are also substrates for their GAP activity. In addition, the role of the Rho family GTPases interacting with plexins remains to be elucidated in more detail. In particular, it remains to be resolved how Rac1 promotes Rnd1 recruitment. In addition, little is known about GTPases interacting with Plexin-C1 and -D1 and it remains to be investigated if all A- and B-type plexins show a comparable behavior in binding GTPases.

While different plexins appear to use a similar mechanism for the regulation of integrin-mediated adhesion, other pathways involved in the response to semaphorins may differ. B-type plexins acitvate Rho through constitutively associated RhoGEFs. The regulation of Rho activity probably is important to modulate actin polymerization. A-type plexins do not contain PDZ-binding motifs and may employ different GEFs and/or different mechanism to influence the actin cytoskeleton or endocytosis.

References

1. Fiore R, Püschel AW. The function of semaphorins during nervous system development. Front Biosci 2003; 8:484-499.
2. He Z, Tessier-Lavigne M. Neuropilin is a receptor for the axonal chemorepellent Semaphorin III. Cell 1997; 90(4):739-751.
3. Kolodkin AL, Levengood DV, Rowe EG et al. Neuropilin is a semaphorin III receptor. Cell 1997; 90(4):753-762.
4. Rohm B, Ottemeyer A, Lohrum M et al. Plexin/neuropilin complexes mediate repulsion by the axonal guidance signal semaphorin 3A. Mech Dev 2000; 93(1-2):95-104.
5. Takahashi T, Fournier A, Nakamura F et al. Plexin-neuropilin-1 complexes form functional semaphorin-3A receptors. Cell 1999; 99(1):59-69.
6. Tamagnone L, Artigiani S, Chen H et al. Plexins are a large family of receptors for transmembrane, secreted, and GPI-anchored semaphorins in vertebrates. Cell 1999; 99(1):71-80.
7. Gu C, Yoshida Y, Livet J et al. Semaphorin 3E and plexin-D1 control vascular pattern independently of neuropilins. Science 2005; 307(5707):265-268.
8. Artigiani S, Conrotto P, Fazzari P et al. Plexin-B3 is a functional receptor for semaphorin 5A. EMBO Rep 2004; 5(7):710-714.
9. Toyofuku T, Zhang H, Kumanogoh A et al. Dual roles of Sema6D in cardiac morphogenesis through region-specific association of its receptor, Plexin-A1, with off-track and vascular endothelial growth factor receptor type 2. Genes Dev 2004; 18(4):435-447.
10. Fan J, Mansfield SG, Redmond T et al. The organization of F-actin and microtubules in growth cones exposed to a brain-derived collapsing factor. J Cell Biol 1993; 121(4):867-878.
11. Fan J, Raper JA. Localized collapsing cues can steer growth cones without inducing their full collapse. Neuron 1995; 14(2):263-274.
12. Mikule K, Gatlin JC, de la Houssaye BA et al. Growth cone collapse induced by semaphorin 3A requires 12/15-lipoxygenase. J Neurosci 2002; 22(12):4932-4941.
13. Serini G, Valdembri D, Zanivan S et al. Class 3 semaphorins control vascular morphogenesis by inhibiting integrin function. Nature 2003; 424(6947):391-397.
14. Castellani V, Falk J, Rougon G. Semaphorin3A-induced receptor endocytosis during axon guidance responses is mediated by L1 CAM. Mol Cell Neurosci 2004; 26(1):89-100.
15. Jurney WM, Gallo G, Letourneau PC et al. Rac1-mediated endocytosis during ephrin-A2- and semaphorin 3A-induced growth cone collapse. J Neurosci 2002; 22(14):6019-6028.
16. Fournier AE, Nakamura F, Kawamoto S et al. Semaphorin3A enhances endocytosis at sites of receptor-F-actin colocalization during growth cone collapse. J Cell Biol 2000; 149(2):411-422.
17. Bagnard D, Sainturet N, Meyronet D et al. Differential MAP kinases activation during semaphorin3A-induced repulsion or apoptosis of neural progenitor cells. Mol Cell Neurosci 2004; 25(4):722-731.
18. Basile JR, Afkhami T, Gutkind JS. Semaphorin 4D/plexin-B1 induces endothelial cell migration through the activation of PYK2, Src, and the phosphatidylinositol 3-kinase-Akt pathway. Mol Cell Biol 2005; 25(16):6889-6898.
19. Brown M, Jacobs T, Eickholt B et al. Alpha2-chimaerin, cyclin-dependent Kinase 5/p35, and its target collapsin response mediator protein-2 are essential components in semaphorin 3A-induced growth-cone collapse. J Neurosci 2004; 24(41):8994-9004.
20. Campbell DS, Holt CE. Apoptotic pathway and MAPKs differentially regulate chemotropic responses of retinal growth cones. Neuron 2003; 37(6):939-952.
21. Eickholt BJ, Walsh FS, Doherty P. An inactive pool of GSK-3 at the leading edge of growth cones is implicated in Semaphorin 3A signaling. J Cell Biol 2002; 157(2):211-217.
22. Fujioka S, Masuda K, Toguchi M et al. Neurotrophic effect of Semaphorin 4D in PC12 cells. Biochem Biophys Res Commun 2003; 301(2):304-310.

23. Mitsui N, Inatome R, Takahashi S et al. Involvement of Fes/Fps tyrosine kinase in semaphorin3A signaling. EMBO J 2002; 21:3274-3285.
24. Pasterkamp RJ, Peschon JJ, Spriggs MK et al. Semaphorin 7A promotes axon outgrowth through integrins and MAPKs. Nature 2003; 424(6947):398-405.
25. Sasaki Y, Cheng C, Uchida Y et al. Fyn and Cdk5 mediate semaphorin-3A signaling, which is involved in regulation of dendrite orientation in cerebral cortex. Neuron 2002; 35(5):907-920.
26. Schwamborn JC, Fiore R, Bagnard D et al. Semaphorin 3A stimulates neurite extension and regulates gene expression in PC12 cells. J Biol Chem 2004; 279(30):30923-30926.
27. Uchida Y, Ohshima T, Sasaki Y et al. Semaphorin3A signaling is mediated via sequential Cdk5 and GSK3beta phosphorylation of CRMP2: Implication of common phosphorylating mechanism underlying axon guidance and Alzheimer's disease. Genes Cells 2005; 10(2):165-179.
28. Takahashi T, Strittmatter SM. PlexinA1 autoinhibition by the plexin sema domain. Neuron 2001; 29(2):429-439.
29. Oinuma I, Katoh H, Negishi M. Molecular dissection of the semaphorin 4D receptor plexin-B1-stimulated R-Ras GTPase-activating protein activity and neurite remodeling in hippocampal neurons. J Neurosci 2004; 24(50):11473-11480.
30. Antipenko A, Himanen JP, van Leyen K et al. Structure of the semaphorin-3A receptor binding module. Neuron 2003; 39(4):589-598.
31. Campbell ID, Ginsberg MH. The talin-tail interaction places integrin activation on FERM ground. Trends Biochem Sci 2004; 29(8):429-435.
32. Kim M, Carman CV, Springer TA. Bidirectional transmembrane signaling by cytoplasmic domain separation in integrins. Science 2003; 301(5640):1720-1725.
33. Takagi J, Petre BM, Walz T et al. Global conformational rearrangements in integrin extracellular domains in outside-in and inside-out signaling. Cell 2002; 110(5):599-611.
34. Wennerberg K, Rossman KL, Der CJ. The Ras superfamily at a glance. J Cell Sci 2005; 118(5):843-846.
35. Wennerberg K, Der CJ. Rho-family GTPases: It's not only Rac and Rho (and I like it). J Cell Sci 2004; 117(8):1301-1312.
36. Nobes CD, Lauritzen I, Mattei MG et al. A new member of the Rho family, Rnd1, promotes disassembly of actin filament structures and loss of cell adhesion. J Cell Biol 1998; 141(1):187-197.
37. Govek EE, Newey SE, Van Aelst L. The role of the Rho GTPases in neuronal development. Genes Dev 2005; 19(1):1-49.
38. Chiarugi P, Cirri P. Redox regulation of protein tyrosine phosphatases during receptor tyrosine kinase signal transduction. Trends Biochem Sci 2003; 28(9):509-514.
39. Jalink K, van Corven EJ, Hengeveld T et al. Inhibition of lysophosphatidate- and thrombin-induced neurite retraction and neuronal cell rounding by ADP ribosylation of the small GTP-binding protein Rho. J Cell Biol 1994; 126(3):801-810.
40. Wahl S, Barth H, Ciossek T et al. Ephrin-A5 induces collapse of growth cones by activating Rho and Rho kinase. J Cell Biol 2000; 149(2):263-270.
41. Aurandt J, Vikis HG, Gutkind JS et al. The semaphorin receptor plexin-B1 signals through a direct interaction with the Rho-specific nucleotide exchange factor, LARG. Proc Natl Acad Sci 2002; 99(19):12085-12090.
42. Hirotani M, Ohoka Y, Yamamoto T et al. Interaction of plexin-B1 with PDZ domain-containing Rho guanine nucleotide exchange factors. Biochem Biophys Res Commun 2002; 297(1):32-37.
43. Oinuma I, Katoh H, Harada A et al. Direct interaction of Rnd1 with Plexin-B1 regulates PDZ-RhoGEF-mediated Rho activation by Plexin-B1 and induces cell contraction in COS-7 cells. J Biol Chem 2003; 278(28):25671-25677.
44. Swiercz JM, Kuner R, Behrens J et al. Plexin-B1 directly interacts with PDZ-RhoGEF/LARG to regulate RhoA and growth cone morphology. Neuron 2002; 35(1):51-63.
45. Jin Z, Strittmatter SM. Rac1 mediates collapsin-1-induced growth cone collapse. J Neurosci 1997; 15(17):6256-6563.
46. Kuhn TB, Brown MD, Wilcox CL et al. Myelin and collapsin-1 induce motor neuron growth cone collapse through different pathways: Inhibition of collapse by opposing mutants of rac1. J Neurosci 1999; 19(6):1965-1975.
47. Västrik I, Eickholt BJ, Walsh FS et al. Sema3A-induced growth-cone collapse is mediated by Rac1 amino acids 17-32. Curr Biol 1999; 9(18):991-998.
48. Luo Y, Raible D, Raper JA. Collapsin: A protein in brain that induces the collapse and paralysis of neuronal growth cones. Cell 1993; 75(2):217-227.
49. Oinuma I, Ishikawa Y, Katoh H et al. The Semaphorin 4D receptor Plexin-B1 is a GTPase activating protein for R-Ras. Science 2004; 305(5685):862-865.

50. Toyofuku T, Yoshida J, Sugimoto T et al. FARP2 triggers signals for Sema3A-mediated axonal repulsion. Nat Neurosci 2005; 8(12):1712-1719.
51. Driessens MH, Hu H, Nobes CD et al. Plexin-B semaphorin receptors interact directly with active Rac and regulate the actin cytoskeleton by activating Rho. Curr Biol 2001; 11(5):339-344.
52. Rohm B, Rahim B, Kleiber B et al. The semaphorin 3A receptor may directly regulate the activity of small GTPases. FEBS Lett 2000; 486(1):68-72.
53. Vikis HG, Li W, He Z et al. The semaphorin receptor plexin-B1 specifically interacts with active rac in a ligand-dependent manner. Proc Natl Acad Sci USA 2000; 97(23):12457-12462.
54. Zanata SM, Hovatta I, Rohm B et al. Antagonistic effects of Rnd1 and RhoD GTPases regulate receptor activity in Semaphorin 3A induced cytoskeletal collapse. J Neurosci 2002; 22(2):471-477.
55. Turner LJ, Nicholls S, Hall A. The activity of the plexin-A1 receptor is regulated by rac. J Biol Chem 2004; 279(32):33199-33205.
56. Vikis HG, Li W, Guan KL. The plexin-B1/Rac interaction inhibits PAK activation and enhances Sema4D ligand binding. Genes Dev 2002; 16(7):836-845.
57. Hu H, Marton TF, Goodman CS. Plexin B mediates axon guidance in Drosophila by simultaneously inhibiting active rac and enhancing rhoA signaling. Neuron 2001; 32(1):39-51.
58. Scheffzek K, Ahmadian MR, Kabsch W et al. The Ras-RasGAP complex: Structural basis for GTPase activation and its loss in oncogenic Ras mutants. Science 1997; 277(5324):333-338.
59. Scheffzek K, Ahmadian MR, Wiesmuller L et al. Structural analysis of the GAP-related domain from neurofibromin and its implications. EMBO J 1998; 17(15):4313-4327.
60. Scheffzek K, Ahmadian MR, Wittinghofer A. GTPase-activating proteins: Helping hands to complement an active site. Trends Biochem Sci 1998; 23(7):257-262.
61. Nakamura F, Tanaka M, Takahashi T et al. Neuropilin-1 extracellular domains mediate semaphorin D/III-induced growth cone collapse. Neuron 1998; 21(5):1093-1100.
62. Kinbara K, Goldfinger LE, Hansen M et al. Ras GTPases: Integrins' friends or foes? Nat Rev Mol Cell Biol 2003; 4(10):767-776.
63. Ling K, Doughman RL, Firestone AJ et al. Type I gamma phosphatidylinositol phosphate kinase targets and regulates focal adhesions. Nature 2002; 420(6911):89-93.
64. Di Paolo G, Pellegrini L, Letinic K et al. Recruitment and regulation of phosphatidylinositol phosphate kinase type 1 gamma by the FERM domain of talin. Nature 2002; 420(6911):85-89.
65. Aizawa H, Wakatsuki S, Ishii A et al. Phosphorylation of cofilin by LIM-kinase is necessary for semaphorin 3A-induced growth cone collapse. Nat Neurosci 2001; 4(4):367-373.
66. Pollard TD, Borisy GG. Cellular motility driven by assembly and disassembly of actin filaments. Cell 2003; 112(4):453-465.
67. Perrot V, Vazquez-Prado J, Gutkind JS. Plexin B regulates Rho through the guanine nucleotide exchange factors leukemia-associated Rho GEF (LARG) and PDZ-RhoGEF. J Biol Chem 2002; 277(45):43115-43120.
68. Barberis D, Casazza A, Sordella R et al. p190 Rho-GTPase activating protein associates with plexins and it is required for semaphorin signaling. J Cell Sci 2005; 118(20):4689-4700.
69. Giordano S, Corso S, Conrotto P et al. The semaphorin 4D receptor controls invasive growth by coupling with Met. Nat Cell Biol 2002; 4(9):720-724.
70. Swiercz JM, Kuner R, Offermanns S. Plexin-B1/RhoGEF-mediated RhoA activation involves the receptor tyrosine kinase ErbB-2. J Cell Biol 2004; 165(6):869-880.
71. Dalpe G, Zhang LW, Zheng H et al. Conversion of cell movement responses to Semaphorin-1 and Plexin-1 from attraction to repulsion by lowered levels of specific RAC GTPases in C. elegans. Development 2004; 131(9):2073-2088.

CHAPTER 3

Intracellular Kinases in Semaphorin Signaling

Aminul Ahmed and Britta J. Eickholt*

Abstract

Originally identified as collapse-inducing and repellent proteins for neuronal processes, semaphorins are now implicated in a diverse array of cellular responses, contributing not only to embryonic development, but also to the maintenance of tissue integrity in the adult organism. In addition, semaphorins play a role in the pathological context. Some Semaphorins can act at a distance, facilitating the navigation of cells or axonal process, whilst others evoke responses in a contact-dependent fashion. The intracellular signaling mechanisms employed by the semaphorins are beginning to be determined, and much work in recent years implicates a host of intracellular kinases in mediating Semaphorin function. These include the tyrosine kinase Fyn and the serine/threonine kinases Cdk5, GSK3, MAPK, and LIMK, and the lipid kinase PI3K. What follows is a review of this work with respect to their functions in mediating specific semaphorin-induced responses.

Introduction

The Semaphorins are a large conserved family of secreted and membrane associated proteins. The first identification of a vertebrate Semaphorin (originally called Collapsin,[1]) was driven by questions regarding the molecular properties that control the 'growth cone collapse': a visible retraction of lamellipodia and filopodia that occurs either upon contact of two different types of neurons,[2,3] or in response to membrane preparations derived from tissues known to inhibit axon extension.[4,5] Since these early experiments, semaphorins have been shown to act primarily as negative regulators of growth cone guidance during development—to influence axon steering and fasciculation—but also to mediate attractive responses.[6-11] The Semaphorins are also implicated in a variety of nonneuronal functions including neural crest cell migration,[12-15] cardiac and skeletal development,[16-19] motility and chemotaxis in cancer cells,[20] vascular development[21,22] and immune responses.[23-25] Over 20 Semaphorins have been characterised, all of which possess a 500 amino acid 'Sema' domain. They are highly conserved among species and have been divided into eight classes, of which the most intensively studied—in terms of the identified requirement of specific kinases—are members of the class 3 secreted and class 4 transmembrane Semaphorins.[26]

Phosphorylation is recognised as a fundamental process involved in the regulation of intracellular signaling in response to extracellular signals, and is mediated by a host of intracellular serine/threonine, tyrosine and lipid kinases. The requirement of kinases in Semaphorin signaling extends from the surface of the cell, at the receptor level, to specific intracellular pathways. In this chapter we will focus on the specific, yet divergent, cellular processes that are achieved through modulation of the enzymatic activity of key intracellular kinase pathways (Fig. 1). For example, LIM-Kinase, GSK3 and Cdk5 appear to be key components in

*Corresponding Author: Britta J. Eickholt—MRC Centre for Developmental Neurobiology, King's College London, London SE1 1UL, U.K. Email: Britta.J.Eickholt@kcl.ac.uk

Semaphorins: Receptor and Intracellular Signaling Mechanisms, edited by R. Jeroen Pasterkamp.
©2007 Landes Bioscience and Springer Science+Business Media.

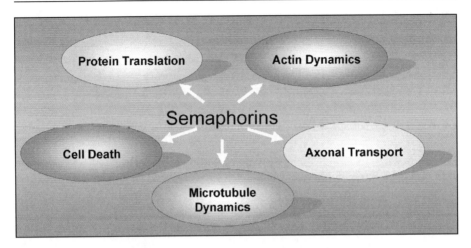

Figure 1. Members of the semaphorin family evoke a number of cellular responses in various cell types. These include changes in actin and microtubule dynamics, axonal transport, protein translation and cell death responses.

the regulation of Semaphorin-induced changes in both actin and microtubule dynamics.[27-30] Several of these kinases—identified by their function in changing microtubule dynamics—also regulate axonal transport.[31] Cdk5 provides a common regulatory mechanism for local protein translation,[31] which also depends on MAP-kinase signaling mediated by p42/p44.[31,32] Consideration of the apoptotic players downstream of its signal further highlights the divergent nature of the Semaphorins. These effects of Semaphorins appear to be mediated by the MAP-kinase p38 and JNK, resulting in selective cell death responses.[33,34] Semaphorin signaling also intersects with the control of cell adhesion mediated by integrins.[35,36] Integrins connect directly to the actin cytoskeleton through their interaction with a number of cytoskeletal proteins, and such aggregates of integrins and cytoskeletal proteins act as scaffolds for the recruitment of different protein kinases. This important Semaphorin function will be discussed in detail in Casazza et al, 2006.

Semaphorins Regulate Actin Filament Dynamics by Controlling Cofilin Activity

Intracellular signaling, mediated by protein kinase cascades, conveys information from the extracellular environment to dynamic changes of actin filament assembly by regulating actin-binding proteins. With regard to Semaphorins, there is compelling evidence for an involvement of members of the ADF/cofilin family in the regulation of actin dynamics.[27] ADF/cofilin proteins are regulated by several protein kinases, including Rho-associated kinase (ROCK) and p21-activated kinase (PAK) (Fig. 2), and their common downstream target LIM-kinase (LIMK, LIM being the acronym of three gene products, Lin-11, Isl-1, and Mec-3). In the Sema3A-mediated growth cone collapse, LIMK dependent inactivation (phosphorylation) of cofilin has been shown to be a necessary signaling component.[27] LIMK dependent phosphorylation of ADF/cofilin leads to its dissociation from actin filament and actin monomers, resulting in reduced actin turnover, decreases in motility and/or a redistribution of actin.[37] Two isoforms of LIMK, LIMK1 and LIMK2 exist; both show a broad expression pattern.[38-40] However, LIMK1 appears to be enriched in neural tissue and has been shown to be present in the growth cone of cultured neurons.[39,41,42] Inhibition of LIMK1, either through expression of a dominant negative protein, or by pharmacological inhibition, has been shown to antagonise the Sema3A induced growth cone collapse.[27] Although evidence of the involvement of LIMK

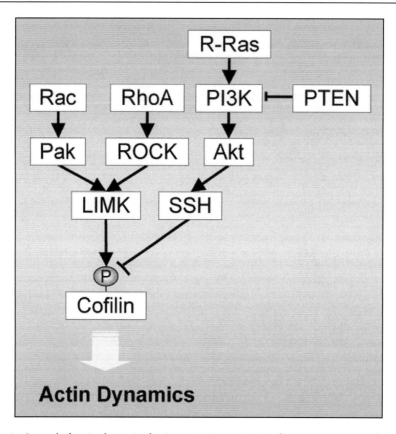

Figure 2. Control of actin dynamics by Sema3A. Sema3A stimulation in neurons induces the transient phosphorylation (inactivation) of the actin binding protein cofilin, which depends on LIM kinase (LIMK) activity. LIMK activation, in turn, may be controlled by members of the p21-activated protein kinases (PAK), activated during Sema3A signaling in a Rac-dependent manner. On the other hand, LIMK-dependent phosphorylation of cofilin during Sema3A stimulation may be induced by RhoA, acting through its target Rho kinase (ROCK). LIMK mediated inactivation of cofilin may be counteracted by activation of cofilin phosphatases, for exmaple, members of the Slingshot (SSH) family. Putative upstream kinases controlling SSH activity during Sema3A stimulation involve phosphatidylinositol 3-kinase (PI3K), phosphatase and tensin homologue deleted on chromosome ten (PTEN) and R-Ras.

in the Sema3A growth cone collapse response is very persuasive, an added complexity has to be considered with regard to regulation of cofilin during this process. For example, although LIMK1 mediated phosphorylation of cofilin is necessary in the Sema3A induced growth inhibition, it was shown not to be sufficient to mimic this response.[27] Similarly, Sema3A evoked in neurons a biphasic response, initially increasing phosphorylation of cofilin, followed by subsequent de-phosphorylation (activation) after 5 minutes.[27] Considering these results in light of known cofilin functions in other cell types,[42-44] a reasonable hypothesis would be that spatiotemporal regulation of cofilin phosphorylation and de-phosphorylation cycles are required in order to appropriately fine-tune neurite outgrowth and growth cone collapse. Thus, LIMK mediated inactivation of cofilin may be counteracted by rapid activation of a phosphatase activity mediated by the family of cofilin phosphatases termed Slingshot (SSH) (Fig. 2). Three mammalian SSH homologues have been identified to date and are reported to be enriched in growth cones.[42]

The idea of a putative involvement of SSH in reactivating cofilin during Sema3A-mediated growth cone collapse is consistent with a similar function reported during Nogo-66 induced axonal growth inhibition.[45] Further studies on the signaling pathways controlling cofilin phosphorylation will be important to clarify the upstream signaling mechanisms that control actin filament dynamics during growth cone behaviour in response to Sema3A. For instance, the exact nature of the upstream kinase regulating LIMK in response to Sema3A remains unclear. Although ROCK could feasibly control LIMK activity during growth cone collapse, p21-activated protein kinases (Pak), an effector of Rac, appears to be the more pertinent upstream kinase[45] (Fig. 2). Pak has been shown to phosphorylate LIMK, which results in increases in phosphorylation of cofilin.[46] In the Sema4D mediated growth cone collapse Plexin-B1 competes with PAK for binding to active Rac in vitro, which inactivates Rac-induced Pak activation.[47] Given that PAK activity regulates the formation of filopodia and membrane ruffles in growth cones,[48] the sequestration of PAK from this complex may results in the collapse of the growth cone, by antagonising LIMK dependent cofilin phosphorylation.[46] It is also interesting to speculate whether SSH is regulated in a PI3K dependent manner during axonal growth inhibition, given that PI3K activity is critical for SSH to elicit activation of cofilin in 293 cells.[49]

Additional studies have identified indispensable functions for cofilin during Semaphorin signaling in other cells. Stimulation of PlexinC1 expressing dendritic cells with the poxvirus A39R, a member of the semaphorin family, has been shown to affect adhesion, which was blocked by a specific inhibitor of cofilin phosphorylation.[50] Similarly, cofilin activation has been reported during Sema3A induced inhibition of agonist-induced actin rearrangement during platelet aggregation.[51]

In summary, downstream effectors of Semaphorin signaling include several kinase enzymatic activities that regulate ADF/cofilin activity, thereby directly controlling the assembly and disassembly of actin filaments. It remains unclear, however, how Sema receptor activation is coupled to these regulatory mechanisms.

Modulation of Semaphorin Responses by Neurotrophins

In neurons, Semaphorin mediated responses in growth cones can be rapidly and differentially modulated by neurotrophins. In one of the earliest studies demonstrating the close relationship between Semaphorin and Neurotrophin signaling, Tuttle and O'Leary found that the Sema3A induced growth cone collapse was more sensitive when the neurons were cultured in brain derived neurotrophic factor (BDNF) as opposed to nerve growth factor (NGF).[52] Moreover, NGF dependent DRG growth cones increased their sensitivity to Sema3A with acute exposure to BDNF. In contrast to this, DRG neurons cultured in NGF concentrations showed an increased growth cone resistance to collapse in response to Sema3A with an acute application of NGF or BDNF.[53] These studies collectively suggest that the responses of growth cones to Semaphorins converge with signaling provided by Neurotrophins.

In the search for the underlying molecular mechanism, Atwal et al demonstrated that Sema3F antagonises NGF-induced activation of the lipid kinase phosphatidylinositol 3-kinase (PI3K) in sympathetic neurons.[54] In addition to this, sustained activation of PI3K—through overexpression of the docking protein growth-associated binder 1 (Gab-1)—was shown to partially reverse Sema3F induced growth cone collapse.[54] PI3Ks are a family of lipid kinases, which regulate a wide variety of biological responses in a broad range of cell types.[55] PI3K activity results in the generation of phosphatidylinositol (3,4,5)-trisphosphate (PIP3), which mediates the recruitment and subsequent activation of several intracellular kinases. The involvement of PIP3 in signaling that determines cell asymmetry in different cell types implies that PI3K signaling can be highly restricted[56] and is exemplified in neurons, where PI3K signaling mediates the determination and maintenance of axons and neurotrophin-mediated axon elongation.[57,58] The picture that emerges in light of these findings suggests that PI3K activity may be a determining event during axonal growth and growth cone collapse, which is a view supported also by our own work. We have recently shown that Sema3A suppresses PI3K signaling in NIE-115 neuroblastoma cells, concomitant with an activation of GSK3, which is an established downstream signaling mediator of the PI3K/Akt pathway in many cell types.[59] In addition, we reported a decreased sensitivity of DRG growth cone to collapse in response to Sema3A by activation of PI3K using a small phosphopeptide known to mimic receptor-dependent activation of PI3K.[59] Not only is inactivation of PI3K in response to Sema3A and Sema3F a requirement for the growth cone collapse, but acute exposure to the PI3K inhibitor LY294002, is sufficient to induce a growth cone collapse in both sympathetic and DRG neurons.[54,59] Thus, Neurotrophin and class III Semaphorins converge to modulate levels of membranous PIP3 and define the fine-tuning between axonal growth and inhibition.

Regulation of PI3K Signaling by Semaphorins

The molecular mechanisms underlying the semaphorin effect in antagonising PI3K signaling remain unidentified. Among the PI3Ks, the class I group of PI3K is key in the generation of the PIP3 signaling intermediary, leading to activation of Akt/PKB and other downstream effectors. Class I PI3Ks are heterodimers consisting of a regulatory subunit (collectively called 'p85s') whose SH2 domain mediates recruitment of the p110 catalytic subunit to membrane-bound receptors, adaptor proteins and Ras.[60,61] Stimulation of the PI3Ks involves binding of p85 to the phosphorylated YXXM motif during growth factor receptor activation.[62,63] Interestingly, the Sema3 receptor PlexinA1 incorporates a putative SH2 binding motif—the characteristic YXXM sequence—raising the possibility of a direct interaction of PlexinA1 with p85 of PI3K. However, we were unable to confirm association of PlexinA1 (or constitutively active PlexinA1Δect) with the PI3K regulatory subunit, p85 (unpublished observation, A.A, B.J.E). Alternatively, Semaphorin downregulation of PI3K signaling may be achieved through inhibition of Ras activity. In this context, Plexins have been reported to contain sequence homology to RasGAPs. For example, PlexinB1 mediated growth cone collapse suppressed R-Ras by activation of its intrinsic RasGAP activity.[64-66] The possibility of R-Ras to control PI3K activity during Sema3A induced growth cone collapse is strengthened by the observation that overexpression of Ras attenuates the PI3K/Akt signaling pathway, resulting in an increase in GSK3 phosphorylation and axonal remodelling in hippocampal neurons.[67] Indeed, downregulation of R-Ras has been reported in response to Sema3A,[64] and a constitutively active R-Ras mutant inhibits Sema3A-mediated growth cone collapse.[65] However, whether Ras-GAP activity controls PI3K/GSK3 signaling in response to Sema3A remains to be demonstrated. Finally, PI3K signaling is terminated by the activities of the lipid phosphatases PTEN (phosphatase and tensin homologue deleted on chromosome ten) and Src-homology 2 (SH2) domain-containing inositol 5'-phosphatase (SHIP),[68,69] and we have recently reported on the involvement of PTEN in the Sema3A mediated growth cone collapse.[59] In DRG neurons, upon Sema3A exposure, transient accumulations of PTEN to the growth cone membrane shifts to a stable recruitment. In addition, PTEN was required for Sema3A mediated changes in PI3K/GSK3 signaling. Considering the recent identification of PTEN as a substrate for GSK3[70]—the hierarchies of PI3K and PTEN in their ability to control PIP3 levels and the potential that GSK3 may provide feedback regulation of PTEN—both require further experimental investigation (see Fig. 3).

Another Semaphorin, Sema4D, has been shown to increase endothelial cell migration, through activation of Akt in a PI3K-dependent manner.[71] Binding of Sema4D to Plexin-B1 was shown to trigger the sequential activation of the nonreceptor tyrosine kinases PYK2 and Src, which recruited PI3K to form a multimeric signaling complex independent of the intrinsic Ras GAP activity of Plexin-B1. This result extends the complexity of PI3K/Akt signaling with regard to Semaphorin function, and suggests that diversity in cellular outcomes of PI3K activity may contribute in a cell type dependent fashion.

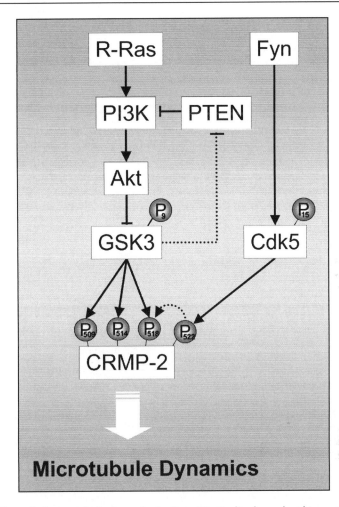

Figure 3. Control of microtubule dynamics by Sema3A. Cyclin dependent kinase 5 (Cdk5) and Glycogen synthase kinase 3 (GSK3) are intracellular mediators of Sema3A in neurons and have been shown to control the activity of collapse response mediator protein-2 (CRMP-2). Cdk5 regulation during Sema3A signaling is subject to phosphorylation at Tyr15 by a member a Src tyrosine kinase, Fyn. Cdk5 directly phosphorylates CRMP-2 at Ser522, which facilitates subsequent phosphorylation of CRMP-2 at Ser518, Thr514 and Thr509 by GSK3, resulting in decreased affinities of CRMP-2 to β-tubulin and changes in microtubule dynamics. During Sema3A induced growth cone collapse, GSK3 activation is mediated through inactivation of phosphatidylinositol 3-kinase (PI3K)/Akt signaling, which requires the lipid phosphatase PTEN. It is interesting to speculate whether GSK3 activity in the Sema3A response also targets its substrate PTEN to quickly adapt and desensitise PI3K/GSK3 signaling. On the other hand, inactivation of PI3K by Sem3A may also depend on inhibition of R-Ras by the intrinsic Plexin RasGAP activity.

Synergistic Control of CRMP-2 Phosphorylation by CDK-5 and GSK3 Mediates Sema3A Function

An established target of PI3K activity in neurons is Glycogen Synthase Kinase 3 (GSK3). GSK3 is a multifunctional serine/threonine kinase originally identified as an enzyme downstream of insulin signaling that is capable of phosphorylating and inactivating glycogen synthase.[72] There are two known isoforms of GSK3 in vertebrates, α and β.[73,74] Highest levels of both isoforms are found in the brain, and the pattern of GSK3β expression broadly correlates with the period of axonal extension and dendritic plasticity.[74,75] Localised PI3K activation in the neuronal growth cone reduces GSK3 activity towards microtubule-binding proteins, which results in increases in axonal extension.[58] Inactivation of GSK3 is achieved by phosphorylation of Ser9 and Ser21 in GSK3β and GSK3α respectively, and our own work characterised an inactive (phosphorylated) pool of the enzyme specifically maintained at the leading edge of motile growth cones.[28] We also demonstrated that inhibition of axonal extension induced by Sema3A leads to GSK3 activation.[28] This response may be achieved, at least in part, through alterations in microtubule dynamics by GSK3 dependent inactivation of collapse response mediator protein-2 (CRMP-2)[29,30] (Fig. 3). Although it is discussed in detail in Schmidt and Strittmatter, 2006, it is necessary to outline key aspects of the convergence of a number of kinases, which directly phosphorylate and regulate the activity of different members of the CRMP family.

5 CRMP isoforms have been characterised to date. CRMP-1, 2 and 4 have been shown to be targeted by GSK3 activity.[29,76,77] Initial experiments by Goshima and colleagues identified CRMP-2 as an essential component of Sema3A signaling. Expression of CRMP-2 in Xenopus oocytes, followed by the application of Sema3A, resulted in an inward current of chloride ions, which originally suggested that CRMP-2 linked Sema3A signaling to intracellular calcium mobilisation.[78] In DRG neurons, antibodies directed against CRMP-2 had no effect on growth cone extension, but antagonised the growth cone collapse induced by Sema3A application.[78] In response to Sema3A, GSK3 mediates CRMP-2 phosphorylation at Ser518/Thr514/Thr509, dependent on a preceding priming of the Ser522 site mediated by a second kinase; cyclin-dependent kinase 5 (Cdk5).[29,30,79] Sema3A is a known activator of Cdk5 in neurons and Cdk5's activity towards CRMP-2 is an essential component in the Sema3A growth cone collapse.[29,30,80] In response to Sema3A, Cdk5 regulation is subject to phosphorylation at Tyr15 by a member of the Src family of tyrosine kinases (Fyn) and is dependent on the recruitment of a Fyn/Cdk5 complex to the PlexinA2 receptor.[80] In line with this model, DRG neurons from both Fyn- and Cdk5 deficient mice show decreased sensitivity to Sema3A. Also, phosphorylation of Tau (a Cdk5 substrate) is observed following Sema3A treatment.[80] CRMP-4 has also been shown to be targeted by GSK3 activity, yet no changes in phosphorylation have been reported in response to Sema3A.[79] This surprising finding may underline the sequential phosphorylation of CRMP-2 described above, involving an initial priming activity by Cdk5 that facilitates phosphorylation by GSK3 (Fig. 3).[29,30,79] Therefore, the specificity of GSK3 activity towards its substrate CRMP-2 in the Sema3A-mediated growth cone collapse may, crucially, depend on a coactivated priming kinase.

A further kinase implicated in the mediation of Sema3A signaling by targeting CRMP-2 is the tyrosine kinase Fes/Fps, which is found in a complex with PlexinA1 and CRMP-2.[81] Expression of a Fes kinase mutant into mouse DRG neurons attenuates the Sema3A induced growth cone collapse, suggesting that Fes/Fps is a candidate linking the Sema3A signal to neuronal responses. This may depend on Fes/Fps tyrosine phosphorylation of CRMP-2, however, neither specific sites nor the consequence of CRMP-2 tyrosine phosphorylation have yet been investigated directly.[81] Finally, a third serine/threonine kinase, Rho-kinase, is involved in regulating CRMP-2 through phosphorylation at Thr-555.[82,83] Whilst Sema3A function does not require Rho-kinase dependent phosphorylation of CRMP-2,[82] it is interesting that ephrin-A5 induces phosphorylation of CRMP-2 via Rho-kinase to induce the growth cone collapse.[83]

Cdk5 and Sema3A-Mediated Increases in Axonal Transport

As well as targeting CRMP-2 in the coordination of microtubule dynamics, Cdk5 has been implicated in the control of axonal transport in response to Sema3A.[84] Cdk5 is a member of the cyclin-dependent kinase family and a protein kinase implicated in the control of numerous aspects of both functional and structural plasticity in the developing and adult nervous system.[85] Given the wide spectrum of Cdk-5 targets in cells,[84,86] it seems feasible that Cdk5 may control additional Sema3A-mediated responses. In this context, Cdk5 activity has been closely linked to the function of microtubule motors. For example, Cdk5 activity towards its substrate dynein-interacting protein Nudel (NudE-like) controls dynein-based axonal transport mechanisms.[87] In addition, a regulatory pathway involving Cdk5, protein phosphatase 1 (PP1) and GSK3 has been demonstrated to control kinesin-driven motility in axons.[88] Application of Sema3A to DRG neurons facilitates both anterograde and retrograde axonal transport in neurons.[89,90] The molecular characterisation of the mechanisms controlling this response has just begun, and recent studies indicate an involvement of the Fyn-Cdk5 complex.[31] Sema3A induced increases in axonal transport in DRG neurons is attenuated by, either, the general tyrosine kinase inhibitor lavendustin A, or, the Cdk5 specific inhibitor olomucine.[31] Furthermore, both Fyn$^{-/-}$ and Cdk5$^{-/-}$ DRG neurons exhibit reduced axonal transport induced by Sema3A when compared to control mice.[31] Interestingly, Sema3A-induced facilitation of axonal transport is dependent on localised protein synthesis in the distal axon segment and/or the growth cone, suggesting that both mechanisms may be subject to control by similar signaling events. This hypothesis is supported by two lines of evidence. Firstly, the observation that Fyn and Cdk5 mediate Sema3A-induced activation of the translation initiation factor eIF-4E in growth cones.[31] Secondly, Sema3A induced anterograde transport of specific messenger RNA (mRNA) in regenerating DRG axons shows sensitivity to lavendustin A and olomucine and is therefore likely to be dependent on Fyn and Cdk5.[91] In summary, the Fyn-Cdk5 complex appears to be a key signaling component in the synchronisation of localised activation of protein synthesis with axonal transport mechanisms required for Sema3A-induced responses in neurons.

MAPK Signaling and the Control of Sema3A Induced Translation of Axonal mRNA

The spatial localization of mRNA to sites of increased cytoskeletal dynamics, in addition to mechanims that control their specific translation in a ligand dependent manner, plays an important function in a number of cellular responses including cell motility and chemotaxis.[92-94] Working with the assumption that cells possess the ability to separate different biological tasks in space and time it is not surprising that in neurons, one biological consequence of Semaphorins is the localised synthesis of proteins in the axonal processes and their growth cones.[32,95] The serine/threonine kinase mTor is the central controller in cellular translation. mTOR modulates the activity of two key translational regulators, the ribosomal S6 kinases (S6K1 and S6K2) and the eukaryotic initiation factor 4E (eIF4E). In general, mTor is active in the presence of growth favourable conditions, and acts in order to maintain a robust rate of protein synthesis and ribosome biogenesis important for cellular homeostasis.[96] Various external stimuli regulate mTor activity, which include growth factors, nutrients, the level of cellular energy (AMP/ATP ratio) and stress.[96]

Navigating axons adapt rapidly their repertoire of responses to external stimuli by increasing anterograde transport of individual mRNA in addition to regulating their translation.[91,97-99] A prevailing example is the dependency of protein translation for attractive turning responses mediated by netrin.[32] Similarly, repulsive turning responses mediated by Sema3A have been demonstrated to depend on newly synthesised proteins in the growth cone.[32] In Xenopus spinal neurons, the presence of the protein synthesis inhibitors anisomycin or cycloheximide blocks the Sema3A induced growth cone collapse, and Sema3A induced repellent activities in the turning assay are antagonised in the presence of protein synthesis inhibitors. Rapamycin, but not wortmannin, inhibits Sema3A-induced growth cone collapse and turning, suggesting that

Sema3A induced translation is independent of PI3K signaling;[32] the major pathway regulating mTor activation and protein synthesis.[100] In contrast, MAPK inhibitors antagonised Sema3A-induced phosphorylation of eukaryotic initiation factors eIF-4EBP1, eIF-4E, and Mnk-1, suggesting a dependency on p42/p44 MAPKs in controlling Sema3A induced transactional activation.[101] These results seem to support the idea that translational initiation by mTOR involves two downstream effectors, p70S6Ks and 4EBPs, which both require phosphorylation by mTOR and MAPK for optimal activation of protein synthesis.[102,103] Because rapid increases in MAPK phosphorylation have been reported as occurring, not only during Sema3A-, but also during Sema3F, Sema7A and Sema4D stimulation,[11,34,54,104] it will be interesting to test whether regulation of protein synthesis is a common target of MAPK activation in response to Semaphorins.

Semaphorin Signaling Leading to Selective Cell Death Responses

In addition to their guidance properties, Semaphorins can also induce cell death. For example, several class III Semaphorins have been implicated in the control of programmed cell death in sensory neurons,[105] small-cell lung cancer cell lines[106,107] and DEV cells.[108] Exposure to exogenous Sema3A in vitro for a prolonged time induces apoptosis of sympathetic and cerebellar granular neurons.[109] Similarly, activated T cells have been shown to trigger neural cell apoptosis by releasing Sema4D.[110] This might reflect a general feature of inhibitory axon guidance molecules; that they may trigger cell death of neurons to eliminate misguided axons and to prevent ectopic, or exuberant innervations of neuronal targets.[105] Little is known about the intracellular kinases ensuring activation of cell death responses by Semaphorins. A partial inhibition of apoptosis has been shown to result from the use of the pan-caspase inhibitor.[105,108] Selective recruitment of the ERK1/2 pathway is also known to occur during Sema3A-induced neural progenitor cell repulsion, whereas p38 MAPK activation is necessary for induction of apoptosis.[34] Recently, Sema3A has been shown to activate c-Jun N-terminal kinase (JNK)/c-Jun signaling, in NGF-dependent DRG neurons. Similarly, pharmacological inhibition of this pathway reduced Sema3A-induced apoptosis,[33] suggesting that activation of the JNK/c-Jun signaling pathway may play an important role in mediating Sema3A-induced cell death responses.

Concluding Remarks

In the 13 years since the identification of the first vertebrate Semaphorin much progress has been made in the identification of the receptors mediating the transduction of Semaphorin signaling. This chapter has detailed some of the evidence for the intracellular protein kinases that mediate these signals, and has detailed the convergence of several kinases onto CRMP-2, the first identified intracellular protein mediating Semaphorin signaling. No doubt, the interplay between the kinases and other well-established aspects of signaling downstream of Semaphorins will have to be further elucidated.

Several neurological conditions have been linked, directly or indirectly, to aberrant Sema3A signaling,[111-118] and studies are beginning to determine the mechanisms of semaphorin function during angiogenesis,[22] during cancer progression[20,119,120] and immune responses.[23,121] The identification and detailed characterisation of the downstream kinases involved opens up the potential for pharmacological intervention by small molecule inhibitors, designed to target single proteins in the signaling network of the cell rather than interfere with global regulators. For instance, in the regeneration of the nervous system, cell permeable analogues have achieved significant recovery following injury.[122-124] These findings are significant, since they indicate that extracellular cues can effectively be bypassed when attempting to postulate strategies for pharmacological promotion of axonal regeneration. It seems clear that the abundance of kinases downstream of Semaphorin will serve as potential therapeutic targets in the treatment of a number of diseases.

References

1. Luo Y, Raible D, Raper JA. Collapsin: A protein in brain that induces the collapse and paralysis of neuronal growth cones. Cell 1993; 75(2):217-227.
2. Kapfhammer JP, Grunewald BE, Raper JA. The selective inhibition of growth cone extension by specific neurites in culture. J Neurosci 1986; 6(9):2527-2534.
3. Kapfhammer JP, Raper JA. Collapse of growth cone structure on contact with specific neurites in culture. J Neurosci 1987; 7(1):201-212.
4. Muller B, Stahl B, Bonhoeffer F. In vitro experiments on axonal guidance and growth-cone collapse. J Exp Biol 1990; 153:29-46.
5. Raper JA, Kapfhammer JP. The enrichment of a neuronal growth cone collapsing activity from embryonic chick brain. Neuron 1990; 4(1):21-29.
6. Polleux F, Morrow T, Ghosh A. Semaphorin 3A is a chemoattractant for cortical apical dendrites. Nature 2000; 404(6778):567-573.
7. Falk J, Bechara A, Fiore R et al. Dual functional activity of semaphorin 3B is required for positioning the anterior commissure. Neuron 2005; 48(1):63-75.
8. Kantor DB, Chivatakarn O, Peer KL et al. Semaphorin 5A is a bifunctional axon guidance cue regulated by heparan and chondroitin sulfate proteoglycans. Neuron 2004; 44(6):961-975.
9. Wolman MA, Liu Y, Tawarayama H et al. Repulsion and attraction of axons by semaphorin3D are mediated by different neuropilins in vivo. J Neurosci 2004; 24(39):8428-8435.
10. Song H, Ming G, He Z et al. Conversion of neuronal growth cone responses from repulsion to attraction by cyclic nucleotides. Science 1998; 281(5382):1515-1518.
11. Pasterkamp RJ, Peschon JJ, Spriggs MK et al. Semaphorin 7A promotes axon outgrowth through integrins and MAPKs. Nature 2003; 424(6947):398-405.
12. Yu HH, Moens CB. Semaphorin signaling guides cranial neural crest cell migration in zebrafish. Dev Biol 2005; 280(2):373-385.
13. Eickholt BJ, Mackenzie SL, Graham A et al. Evidence for collapsin-1 functioning in the control of neural crest migration in both trunk and hindbrain regions. Development 1999; 126(10):2181-2189.
14. Osborne NJ, Begbie J, Chilton JK et al. Semaphorin/neuropilin signaling influences the positioning of migratory neural crest cells within the hindbrain region of the chick. Dev Dyn 2005; 232(4):939-949.
15. Gammill LS, Gonzalez C, Gu C et al. Guidance of trunk neural crest migration requires neuropilin 2/semaphorin 3F signaling. Development 2006; 133(1):99-106.
16. Gitler AD, Lu MM, Epstein JA. PlexinD1 and semaphorin signaling are required in endothelial cells for cardiovascular development. Dev Cell 2004; 7(1):107-116.
17. Brown CB, Feiner L, Lu MM et al. PlexinA2 and semaphorin signaling during cardiac neural crest development. Development 2001; 128(16):3071-3080.
18. Toyofuku T, Zhang H, Kumanogoh A et al. Dual roles of Sema6D in cardiac morphogenesis through region-specific association of its receptor, Plexin-A1, with off-track and vascular endothelial growth factor receptor type 2. Genes Dev 2004; 18(4):435-447.
19. Toyofuku T, Zhang H, Kumanogoh A et al. Guidance of myocardial patterning in cardiac development by Sema6D reverse signaling. Nat Cell Biol 2004; 6(12):1204-1211.
20. Guttmann-Raviv N, Kessler O, Shraga-Heled N et al. The neuropilins and their role in tumorigenesis and tumor progression. Cancer Letters 2006; 231(1):1-11.
21. Deutsch U. Semaphorins guide PerPlexeD endothelial cells. Dev Cell 2004; 7(1):1-2.
22. Klagsbrun M, Eichmann A. A role for axon guidance receptors and ligands in blood vessel development and tumor angiogenesis. Cytokine Growth Factor Rev 2005; 16(4-5):535-548.
23. Takegahara N, Kumanogoh A, Kikutani H. Semaphorins: A new class of immunoregulatory molecules. Philos Trans R Soc Lond B Biol Sci 2005; 360(1461):1673-1680.
24. Kumanogoh A, Shikina T, Watanabe C et al. Requirement for CD100-CD72 interactions in fine-tuning of B-cell antigen receptor signaling and homeostatic maintenance of the B-cell compartment. Int Immunol 2005; 17(10):1277-1282.
25. Kumanogoh A, Shikina T, Suzuki K et al. Nonredundant roles of Sema4A in the immune system: Defective T cell priming and Th1/Th2 regulation in Sema4A-deficient mice. Immunity 2005; 22(3):305-316.
26. Goodman CS, Kolodkin AL, Luo Y et al. Unified nomenclature for the semaphorins/collapsins. Cell 1999; 97(5):551-552.
27. Aizawa H, Wakatsuki S, Ishii A et al. Phosphorylation of cofilin by LIM-kinase is necessary for semaphorin 3A-induced growth cone collapse. Nat Neurosci 2001; 4(4):367-373.
28. Eickholt BJ, Walsh FS, Doherty P. An inactive pool of GSK-3 at the leading edge of growth cones is implicated in Semaphorin 3A signaling. J Cell Biol 2002; 157(2):211-217.

29. Brown M, Jacobs T, Eickholt B et al. Alpha2-chimaerin, cyclin-dependent Kinase 5/p35, and its target collapsin response mediator protein-2 are essential components in semaphorin 3A-induced growth-cone collapse. J Neurosci 2004; 24(41):8994-9004.
30. Uchida Y, Ohshima T, Sasaki Y et al. Semaphorin3A signaling is mediated via sequential Cdk5 and GSK3beta phosphorylation of CRMP2: Implication of common phosphorylating mechanism underlying axon guidance and Alzheimer's disease. Genes Cells 2005; 10(2):165-179.
31. Li C, Sasaki Y, Takei K et al. Correlation between semaphorin3A-induced facilitation of axonal transport and local activation of a translation initiation factor eukaryotic translation initiation factor 4E. J Neurosci 2004; 24(27):6161-6170.
32. Campbell DS, Holt CE. Chemotropic responses of retinal growth cones mediated by rapid local protein synthesis and degradation. Neuron 2001; 32(6):1013-1026.
33. Ben-Zvi A, Yagil Z, Hagalili Y et al. Semaphorin 3A and neurotrophins: A balance between apoptosis and survival signaling in embryonic DRG neurons. J Neurochem 2006; 96(2):585-597.
34. Bagnard D, Sainturet N, Meyronet D et al. Differential MAP kinases activation during semaphorin3A-induced repulsion or apoptosis of neural progenitor cells. Mol Cell Neurosci 2004; 25(4):722-731.
35. Barberis D, Artigiani S, Casazza A et al. Plexin signaling hampers integrin-based adhesion, leading to Rho-kinase independent cell rounding, and inhibiting lamellipodia extension and cell motility. Faseb J 2004; 18(3):592-594.
36. Barberis D, Casazza A, Sordella R et al. p190 Rho-GTPase activating protein associates with plexins and it is required for semaphorin signaling. J Cell Sci 2005; 118(Pt 20):4689-4700.
37. Bamburg JR, Wiggan OP. ADF/cofilin and actin dynamics in disease. Trends Cell Biol 2002; 12(12):598-605.
38. Proschel C, Blouin MJ, Gutowski NJ et al. Limk1 is predominantly expressed in neural tissues and phosphorylates serine, threonine and tyrosine residues in vitro. Oncogene 1995; 11(7):1271-1281.
39. Foletta VC, Moussi N, Sarmiere PD et al. LIM kinase 1, a key regulator of actin dynamics, is widely expressed in embryonic and adult tissues. Exp Cell Res 2004; 294(2):392-405.
40. Okano I, Hiraoka J, Otera H et al. Identification and characterization of a novel family of serine/threonine kinases containing two N-terminal LIM motifs. J Biol Chem 1995; 270(52):31321-31330.
41. Endo M, Ohashi K, Sasaki Y et al. Control of growth cone motility and morphology by LIM kinase and Slingshot via phosphorylation and dephosphorylation of cofilin. J Neurosci 2003; 23(7):2527-2537.
42. Sarmiere PD, Bamburg JR. Regulation of the neuronal actin cytoskeleton by ADF/cofilin. J Neurobiol 2004; 58(1):103-117.
43. Soosairajah J, Maiti S, Wiggan O et al. Interplay between components of a novel LIM kinase-slingshot phosphatase complex regulates cofilin. EMBO J 2005; 24(3):473-486.
44. Nishita M, Tomizawa C, Yamamoto M et al. Spatial and temporal regulation of cofilin activity by LIM kinase and Slingshot is critical for directional cell migration. J Cell Biol 2005; 171(2):349-359.
45. Hsieh SH, Ferraro GB, Fournier AE. Myelin-associated inhibitors regulate cofilin phosphorylation and neuronal inhibition through LIM kinase and Slingshot phosphatase. J Neurosci 2006; 26(3):1006-1015.
46. Edwards DC, Sanders LC, Bokoch GM et al. Activation of LIM-kinase by Pak1 couples Rac/Cdc42 GTPase signaling to actin cytoskeletal dynamics. Nat Cell Biol 1999; 1(5):253-259.
47. Vikis HG, Li W, Guan KL. The plexin-B1/Rac interaction inhibits PAK activation and enhances Sema4D ligand binding. Genes Dev 2002; 16(7):836-845.
48. Robles E, Woo S, Gomez TM. Src-dependent tyrosine phosphorylation at the tips of growth cone filopodia promotes extension. J Neurosci 2005; 25(33):7669-7681.
49. Nishita M, Wang Y, Tomizawa C et al. Phosphoinositide 3-kinase-mediated activation of cofilin phosphatase Slingshot and its role for insulin-induced membrane protrusion. J Biol Chem 2004; 279(8):7193-7198.
50. Walzer T, Galibert L, Comeau MR et al. Plexin C1 engagement on mouse dendritic cells by viral semaphorin A39R induces actin cytoskeleton rearrangement and inhibits integrin-mediated adhesion and chemokine-induced migration. J Immunol 2005; 174(1):51-59.
51. Kashiwagi H, Shiraga M, Kato H et al. Negative regulation of platelet function by a secreted cell repulsive protein, semaphorin 3A. Blood 2005; 106(3):913-921.
52. Tuttle R, O'Leary DD. Neurotrophins rapidly modulate growth cone response to the axon guidance molecule, collapsin-1. Mol Cell Neurosci 1998; 11(1-2):1-8.
53. Dontchev VD, Letourneau PC. Nerve growth factor and semaphorin 3A signaling pathways interact in regulating sensory neuronal growth cone motility. J Neurosci 2002; 22(15):6659-6669.

54. Atwal JK, Singh KK, Tessier-Lavigne M et al. Semaphorin 3F antagonizes neurotrophin-induced phosphatidylinositol 3-kinase and mitogen-activated protein kinase kinase signaling: A mechanism for growth cone collapse. J Neurosci 2003; 23(20):7602-7609.
55. Cantley LC. The phosphoinositide 3-kinase pathway. Science 2002; 296(5573):1655-1657.
56. Fivaz M, Meyer T. Specific localization and timing in neuronal signal transduction mediated by protein-lipid interactions. Neuron 2003; 40(2):319-330.
57. Shi SH, Jan LY, Jan YN. Hippocampal neuronal polarity specified by spatially localized mPar3/mPar6 and PI 3-kinase activity. Cell 2003; 112(1):63-75.
58. Zhou FQ, Zhou J, Dedhar S et al. NGF-induced axon growth is mediated by localized inactivation of GSK-3beta and functions of the microtubule plus end binding protein APC. Neuron 2004; 42(6):897-912.
59. Chadborn NH, Ahmed AI, Holt MR et al. PTEN couples Sema3A signaling to growth cone collapse. J Cell Sci 2006; 119(Pt 5):951-957.
60. Escobedo JA, Kaplan DR, Kavanaugh WM et al. A phosphatidylinositol-3 kinase binds to platelet-derived growth factor receptors through a specific receptor sequence containing phosphotyrosine. Mol Cell Biol 1991; 11(2):1125-1132.
61. Carpenter CL, Cantley LC. Phosphoinositide kinases. Biochemistry 1990; 29(51):11147-11156.
62. Wu H, Windmiller DA, Wang L et al. YXXM motifs in the PDGF-beta receptor serve dual roles as phosphoinositide 3-kinase binding motifs and tyrosine-based endocytic sorting signals. J Biol Chem 2003; 278(42):40425-40428.
63. Kapeller R, Toker A, Cantley LC et al. Phosphoinositide 3-kinase binds constitutively to alpha/beta-tubulin and binds to gamma-tubulin in response to insulin. J Biol Chem 1995; 270(43):25985-25991.
64. Toyofuku T, Yoshida J, Sugimoto T et al. FARP2 triggers signals for Sema3A-mediated axonal repulsion. Nat Neurosci 2005; 8(12):1712-1719.
65. Oinuma I, Ishikawa Y, Katoh H et al. The Semaphorin 4D receptor Plexin-B1 is a GTPase activating protein for R-Ras. Science 2004; 305(5685):862-865.
66. Oinuma I, Katoh H, Negishi M. Molecular dissection of the semaphorin 4D receptor plexin-B1-stimulated R-Ras GTPase-activating protein activity and neurite remodeling in hippocampal neurons. J Neurosci 2004; 24(50):11473-11480.
67. Yoshimura T, Arimura N, Kawano Y et al. Ras regulates neuronal polarity via the PI3-kinase/Akt/GSK-3beta/CRMP-2 pathway. Biochem Biophys Res Commun 2006; 340(1):62-68.
68. Leslie NR, Yang X, Downes CP et al. The regulation of cell migration by PTEN. Biochem Soc Trans 2005; 33(Pt 6):1507-1508.
69. Kalesnikoff J, Sly LM, Hughes MR et al. The role of SHIP in cytokine-induced signaling. Rev Physiol Biochem Pharmacol 2003; 149:87-103.
70. Al-Khouri AM, Ma Y, Togo SH et al. Cooperative phosphorylation of the tumor suppressor phosphatase and tensin homologue (PTEN) by casein kinases and glycogen synthase kinase 3beta. J Biol Chem 2005; 280(42):35195-35202.
71. Basile JR, Afkhami T, Gutkind JS. Semaphorin 4D/plexin-B1 induces endothelial cell migration through the activation of PYK2, Src, and the phosphatidylinositol 3-kinase-Akt pathway. Mol Cell Biol 2005; 25(16):6889-6898.
72. Rylatt DB, Aitken A, Bilham T et al. Glycogen synthase from rabbit skeletal muscle. Amino acid sequence at the sites phosphorylated by glycogen synthase kinase-3, and extension of the N-terminal sequence containing the site phosphorylated by phosphorylase kinase. Eur J Biochem 1980; 107(2):529-537.
73. Woodgett JR. Molecular cloning and expression of glycogen synthase kinase-3/factor A. EMBO J 1990; 9(8):2431-2438.
74. Plyte SE, Hughes K, Nikolakaki E et al. Glycogen synthase kinase-3: Functions in oncogenesis and development. Biochim Biophys Acta 1992; 1114(2-3):147-162.
75. Leroy K, Brion JP. Developmental expression and localization of glycogen synthase kinase-3beta in rat brain. J Chem Neuroanat 1999; 16(4):279-293.
76. Cole AR, Knebel A, Morrice NA et al. GSK-3 phosphorylation of the Alzheimer epitope within collapsin response mediator proteins regulates axon elongation in primary neurons. J Biol Chem 2004; 279(48):50176-50180.
77. Yoshimura T, Kawano Y, Arimura N et al. GSK-3beta regulates phosphorylation of CRMP-2 and neuronal polarity. Cell 2005; 120(1):137-149.
78. Goshima Y, Nakamura F, Strittmatter P et al. Collapsin-induced growth cone collapse mediated by an intracellular protein related to UNC-33. Nature 1995; 376(6540):509-514.

79. Cole AR, Causeret F, Yagirdi G et al. Distinct priming kinases contribute to differential regulation of CRMP isoforms by GSK3 in vivo. J Biol Chem 279(48):50176-50180.
80. Sasaki Y, Cheng C, Uchida Y et al. Fyn and Cdk5 mediate semaphorin-3A signaling, which is involved in regulation of dendrite orientation in cerebral cortex. Neuron 2002; 35(5):907-920.
81. Mitsui N, Inatome R, Takahashi S et al. Involvement of Fes/Fps tyrosine kinase in semaphorin3A signaling. EMBO J 2002; 21(13):3274-3285.
82. Arimura N, Inagaki N, Chihara K et al. Phosphorylation of collapsin response mediator protein-2 by Rho-kinase. Evidence for two separate signaling pathways for growth cone collapse. J Biol Chem 2000; 275(31):23973-23980.
83. Arimura N, Menager C, Kawano Y et al. Phosphorylation by Rho kinase regulates CRMP-2 activity in growth cones. Mol Cell Biol 2005; 25(22):9973-9984.
84. Smith DS, Tsai LH. Cdk5 behind the wheel: A role in trafficking and transport? Trends Cell Biol 2002; 12(1):28-36.
85. Nikolic M. The molecular mystery of neuronal migration: FAK and Cdk5. Trends Cell Biol 2004; 14(1):1-5.
86. Cruz JC, Tsai LH. A Jekyll and Hyde kinase: Roles for Cdk5 in brain development and disease. Curr Opin Neurobiol 2004; 14(3):390-394.
87. Niethammer M, Smith DS, Ayala R et al. NUDEL is a novel Cdk5 substrate that associates with LIS1 and cytoplasmic dynein. Neuron 2000; 28(3):697-711.
88. Morfini G, Szebenyi G, Brown H et al. A novel CDK5-dependent pathway for regulating GSK3 activity and kinesin-driven motility in neurons. EMBO J 2004; 23(11):2235-2245.
89. Goshima Y, Hori H, Sasaki Y et al. Growth cone neuropilin-1 mediates collapsin-1/Sema III facilitation of antero- and retrograde axoplasmic transport. J Neurobiol 1999; 39(4):579-589.
90. Goshima Y, Kawakami T, Hori H et al. A novel action of collapsin: Collapsin-1 increases antero- and retrograde axoplasmic transport independently of growth cone collapse. J Neurobiol 1997; 33(3):316-328.
91. Willis D, van Niekerk E, Merianda TT et al. Axonal stimuli specifically regulate anterograde transport of individual mRNAs.
92. Huttelmaier S, Zenklusen D, Lederer M et al. Spatial regulation of beta-actin translation by Src-dependent phosphorylation of ZBP1. Nature 2005; 438(7067):512-515.
93. Shestakova EA, Singer RH, Condeelis J. The physiological significance of beta -actin mRNA localization in determining cell polarity and directional motility. Proc Natl Acad Sci USA 2001; 98(13):7045-7050.
94. Lawrence JB, Singer RH. Intracellular localization of messenger RNAs for cytoskeletal proteins. Cell 1986; 45(3):407-415.
95. Willis DE, Twiss JL. The evolving roles of axonally synthesized proteins in regeneration. Curr Opin Neurobiol 2006; 16(1):111-118.
96. Wullschleger S, Loewith R, Hall MN. TOR signaling in growth and metabolism. Cell 2006; 124(3):471-484.
97. Piper M, Anderson R, Dwivedy A et al. Signaling mechanisms underlying Slit2-induced collapse of Xenopus retinal growth cones. Neuron 2006; 49(2):215-228.
98. Piper M, Holt C. RNA translation in axons. Annu Rev Cell Dev Biol 2004; 20:505-523.
99. Verma P, Chierzi S, Codd AM et al. Axonal protein synthesis and degradation are necessary for efficient growth cone regeneration. J Neurosci 2005; 25(2):331-342.
100. Guertin DA, Sabatini DM. An expanding role for mTOR in cancer. Trends Mol Med 2005; 11(8):353-361.
101. Campbell DS, Holt CE. Apoptotic pathway and MAPKs differentially regulate chemotropic responses of retinal growth cones. Neuron 2003; 37(6):939-952.
102. Schmelzle T, Hall MN. TOR, a central controller of cell growth. Cell 2000; 103(2):253-262.
103. Herbert TP, Tee AR, Proud CG. The extracellular signal-regulated kinase pathway regulates the phosphorylation of 4E-BP1 at multiple sites. J Biol Chem 2002; 277(13):11591-11596.
104. Aurandt J, Li W, Guan KL. Semaphorin 4D activates the MAPK pathway downstream of plexin-B1. Biochem J 2006; 394(Pt 2):459-464.
105. Gagliardini V, Fankhauser C. Semaphorin III can induce death in sensory neurons. Mol Cell Neurosci 1999; 14(4-5):301-316.
106. Tomizawa Y, Sekido Y, Kondo M et al. Inhibition of lung cancer cell growth and induction of apoptosis after reexpression of 3p21.3 candidate tumor suppressor gene SEMA3B. Proc Natl Acad Sci USA 2001; 98(24):13954-13959.
107. Ochi K, Mori T, Toyama Y et al. Identification of semaphorin3B as a direct target of p53. Neoplasia 2002; 4(1):82-87.

108. Bagnard D, Vaillant C, Khuth ST et al. Semaphorin 3A-vascular endothelial growth factor-165 balance mediates migration and apoptosis of neural progenitor cells by the recruitment of shared receptor. J Neurosci 2001; 21(10):3332-3341.
109. Shirvan A, Ziv I, Fleminger G et al. Semaphorins as mediators of neuronal apoptosis. J Neurochem 1999; 73(3):961-971.
110. Giraudon P, Vincent P, Vuaillat C et al. Semaphorin CD100 from activated T lymphocytes induces process extension collapse in oligodendrocytes and death of immature neural cells. J Immunol 2004; 172(2):1246-1255.
111. de Winter F, Cui Q, Symons N et al. Expression of class-3 semaphorins and their receptors in the neonatal and adult rat retina. Invest Ophthalmol Vis Sci 2004; 45(12):4554-4562.
112. Niclou SP, Franssen EH, Ehlert EM et al. Meningeal cell-derived semaphorin 3A inhibits neurite outgrowth. Mol Cell Neurosci 2003; 24(4):902-912.
113. de Wit J, Verhaagen J. Role of semaphorins in the adult nervous system. Prog Neurobiol 2003; 71(2-3):249-267.
114. Scarlato M, Ara J, Bannerman P et al. Induction of neuropilins-1 and -2 and their ligands, Sema3A, Sema3F, and VEGF, during Wallerian degeneration in the peripheral nervous system. Exp Neurol 2003; 183(2):489-498.
115. Moreau-Fauvarque C, Kumanogoh A, Camand E et al. The transmembrane semaphorin Sema4D/CD100, an inhibitor of axonal growth, is expressed on oligodendrocytes and upregulated after CNS lesion. J Neurosci 2003; 23(27):9229-9239.
116. Good PF, Alapat D, Hsu A et al. A role for semaphorin 3A signaling in the degeneration of hippocampal neurons during Alzheimer's disease. J Neurochem 2004; 91(3):716-736.
117. Mah S, Nelson MR, Delisi LE et al. Identification of the semaphorin receptor PLXNA2 as a candidate for susceptibility to schizophrenia. Mol Psychiatry 2006.
118. Maraganore DM, de Andrade M, Lesnick TG et al. High-resolution whole-genome association study of Parkinson disease. Am J Hum Genet 2005; 77(5):685-693.
119. Chedotal A, Kerjan G, Moreau-Fauvarque C. The brain within the tumor: New roles for axon guidance molecules in cancers. Cell Death Differ 2005; 12(8):1044-1056.
120. Bachelder RE, Lipscomb EA, Lin X et al. Competing autocrine pathways involving alternative neuropilin-1 ligands regulate chemotaxis of carcinoma cells. Cancer Res 2003; 63(17):5230-5233.
121. Kumanogoh A, Kikutani H. Roles of the semaphorin family in immune regulation. Adv Immunol 2003; 81:173-198.
122. Qiu J, Cai D, Dai H et al. Spinal axon regeneration induced by elevation of cyclic AMP. Neuron 2002; 34(6):895-903.
123. Neumann S, Bradke F, Tessier-Lavigne M et al. Regeneration of sensory axons within the injured spinal cord induced by intraganglionic cAMP elevation. Neuron 2002; 34(6):885-893.
124. Fournier AE, Takizawa BT, Strittmatter SM. Rho kinase inhibition enhances axonal regeneration in the injured CNS. J Neurosci 2003; 23(4):1416-1423.

CHAPTER 4

MICAL Flavoprotein Monooxygenases:
Structure, Function and Role in Semaphorin Signaling

Sharon M. Kolk and R. Jeroen Pasterkamp*

Abstract

MICALs (for *M*olecule *I*nteracting with *CasL*) form a recently discovered family of evolutionary conserved signal transduction proteins. They contain multiple well-conserved domains known for interactions with the cytoskeleton, cytoskeletal adaptor proteins, and other signaling proteins. In addition to their ability to bind other proteins, MICALs contain a large NADPH-dependent flavoprotein monooxygenase enzymatic domain. Although MICALs have already been implicated in a variety of cellular processes, their function during axonal pathfinding in the *Drosophila* neuromuscular system has been best characterized. During the establishment of neuromuscular connectivity in the fruit fly, MICAL binds the axon guidance receptor Plexin A and transduces semaphorin-1a-mediated repulsive axon guidance. Intriguingly, mutagenesis and pharmacological inhibitor studies suggest a role for MICAL flavoenzyme redox functions in semaphorin/plexin-mediated axonal pathfinding events. This review summarizes our current understanding of MICALs, with an emphasis on their role in semaphorin signaling.

Introduction

The formation of neuronal circuits during embryonic development relies upon the guidance of growing axons to their synaptic targets. To help neurons find these targets, axons are tipped with a highly motile sensory structure called the growth cone. Growth cones are instructed to follow predetermined trajectories by heterogeneously distributed guidance molecules in their extracellular environment. Binding of axon guidance molecules to receptor complexes on the growth cone surface initiates intracellular signaling events, which in turn modulate growth cone morphology and directional motility through local modifications of the neuronal cytoskeleton. Axon guidance molecules can act as attractants or repellents, either directing growth cones towards a specific structure or preventing them from entering inappropriate regions of the embryo. Furthermore, they exist as membrane-associated or soluble agents, acting at short ranges or at long distance, respectively.[1-3]

The semaphorins are among the largest of the conserved families of axon guidance molecules, affecting axon steering, zonal segregation of distinct axon populations, axon branching, axon fasciculation, neuron polarity, and synapse formation and function.[4,5] In addition, semaphorins and their receptors participate in the regulation of various nonneuronal processes such as angiogenesis, organogenesis, tumorigenesis and immune cell function. Furthermore, they may contribute to the onset and/or progression of human (brain) disease.[6-9] Semaphorin

---

*Corresponding Author: R. Jeroen Pasterkamp—Department of Pharmacology and Anatomy, Rudolf Magnus Institute of Neuroscience, University Medical Center Utrecht, Universiteitsweg 100, 3584 CG Utrecht, The Netherlands. Email: j.pasterkamp@med.uu.nl

Semaphorins: Receptor and Intracellular Signaling Mechanisms, edited by R. Jeroen Pasterkamp.
©2007 Landes Bioscience and Springer Science+Business Media.

signaling during axon guidance is dependent on multimeric receptor complexes on the growth cone cell surface that contain plexin proteins as obligatory signal-transducing subunits. Semaphorins can either bind directly to plexins (Sema classes 1, 2, 4-7 and Sema3E) or may require specialized ligand-binding coreceptors, such as neuropilin proteins (Sema class 3), to achieve steering of neuronal processes. In addition to plexins and neuropilins, several structurally unrelated receptors have been identified, some of which function in the nervous system including receptor tyrosine kinases (Otk), Ig superfamily cell adhesion molecules (L1 and NrCAM), integrins, and CD72.[10-12] In sharp contrast to the wealth of information on the biology of semaphorins and their (co)receptors, the cytosolic signaling pathways that mediate growth cone responses to semaphorins are only now beginning to be understood. An ever increasing number of signaling proteins are now implicated in linking semaphorin receptors to the neuronal cytoskeleton including members of the Rho family of small GTPases, collapsin response mediator proteins (CRMPs), and intracellular protein kinases.[11-17] Here we focus on MICALs, a novel family of cytosolic signaling proteins implicated in mediating semaphorin signaling events in neurons. A short description of the recently established neuronal MICAL expression patterns precedes our overview of potential roles played by these multidomain proteins in repulsive semaphorin/plexin signaling. We then review potential roles for MICALs in the regulation of cytoskeletal dynamics and summarize recent insights into the structure and function of the MICAL flavoprotein monooxygenase domain.

The MICAL Family

A critical step in neuronal semaphorin signaling is the activation of specific signal transduction pathways by plexins. Several of the semaphorin signaling cues identified to date can associate directly with the cytoplasmic domain of plexin. The MICAL proteins (for Molecule Interacting with CasL)[18] form a family of cytosolic plexin-interacting proteins that participate in repulsive semaphorin signaling in neurons.[19] In invertebrate species such as *Drosophila*, a single MICAL protein has been identified (D-MICAL), while vertebrates have three *MICAL* genes (*MICAL-1*, *MICAL-2* and *MICAL-3*). In addition, several MICAL-like genes (*MICAL-L*) exist which encode potentially MICAL-related proteins that lack the highly conserved NH_2-terminal region present in 'full-length' MICALs.[19,20] From sequence analysis it has been shown that MICALs contain multiple domains and motifs known to be important for interactions with the actin cytoskeleton and other proteins critical for signaling events to the cytoskeleton. In addition, MICALs uniquely combine their protein-binding properties with an NH_2-terminal flavoprotein monooxygenase domain (Fig. 1A). MICALs are expressed in various tissues including lung, heart, thymus, and brain.[18,19,21,22,60]

MICAL Expression

Thus far, MICAL expression patterns in invertebrate and vertebrate nervous systems have been studied in most detail. In the early embryonic fruit fly (stage 7-8), prominent *D-MICAL* labeling is found in the ventral neurogenic region. At later stages (stage 13 onward), *D-MICAL* transcripts are present within the developing *Drosophila* brain and ventral cord in most, if not all, central nervous system (CNS) neurons. In contrast, mRNA expression in peripheral nervous system (PNS) neurons is weak. In line with D-MICAL's role downstream of Plexin A (PlexA), *D-MICAL* and *PlexA* distribution patterns are highly similar. Immunohistochemistry for D-MICAL protein labels neuronal cell bodies, axons and growth cones.[19]

Vertebrate members of the MICAL family are expressed throughout the developing and mature rat nervous system. In contrast to *MICAL-1* and *-3*, the onset of *MICAL-2* expression is delayed (late embryonic/postnatal) and *MICAL-2* is absent from certain specific brain areas such as the hypothalamus and striatum.[22] *MICAL-1*, *-2* and *-3* expression patterns in the embryonic and postnatal nervous system support the idea that MICALs play roles in neural development, while their presence in adult neurons hints at a possible function in the control of adult (structural) plasticity.[22] Furthermore, the overlap between *MICAL*, *plexin* and *neuropilin* distribution patterns supports a role for vertebrate MICALs in semaphorin signaling.

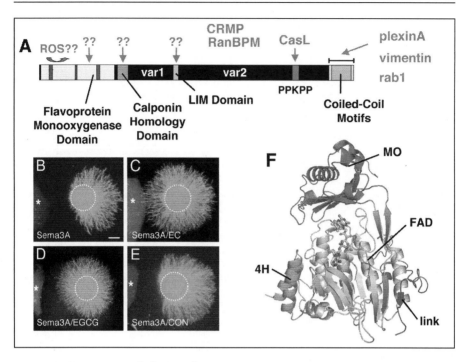

Figure 1. MICALs are multidomain flavoprotein monooxygenases. A) MICALs contain an NH_2-terminal FAD-binding monooxygenase domain of about 500 amino acids followed by a calponin homology domain, a LIM domain, a Pro-Pro-Lys-Pro-Pro (PPKPP) region for Src homology 3 (SH3) recognition, and COOH-terminal coiled-coil motifs. In addition, two variable regions (var1 and var2) are present with no known homology to published sequences. Thus far, MICALs have been reported to associate with CasL, plexinA, vimentin, rab1, CRMPs and RanBPM. The region of the MICAL protein that binds to CRMPs or RanBPM is unknown. Binding partners of other domains (question marks) are unknown. B-E) Flavoprotein monooxygenase inhibitors block Sema3A-induced sensory axon repulsion. E14 rat DRG explants were cocultured with 293 cells expressing Sema3A (asterisks) and grown for 48 h in the presence of vehicle (B) or an inhibitor (C-E). B) Sema3A repels rat sensory neurons. The monooxygenase inhibitors EC (C) and EGCG (D), but not the xanthine oxidase inhibitor allopurinol (CON) (E), inhibit this Sema3A-dependent axon repulsion. For more detail see Terman et al. (2002) and Pasterkamp et al., (2006). F) Ribbon diagram of the mouse MICAL-1 flavoprotein monooxygenase domain. Shown are the four-helix bundle domain (4H), FAD-binding domain (FAD), monooxygenase domain (MO), and linker region (link). The FAD molecule is drawn as balls and sticks. Panels B-E are reprinted from Terman et al. (2002) with permission from Elsevier. ROS, reactive oxygen species.

In addition to being expressed in neurons, *MICAL-1* and *MICAL-3* are expressed in a subset of oligodendrocytes in the postnatal and adult CNS. Several recent studies demonstrate a prominent role for semaphorins and associated signaling pathways in oligodendroglia biology. Oligodendrocyte progenitors and oligodendrocytes express a broad spectrum of semaphorins and semaphorin receptors.[23-28] Furthermore, semaphorins belonging to different subclasses can induce process retraction in oligodendrocyte progenitors and oligodendrocytes, and they can also direct the migration of oligodendrocyte progenitors in vitro.[23,24,26,27,29] Although their function in oligodendrocytes requires further study, MICALs may also act to mediate the effects of semaphorins on oligodendrocyte morphology. This is especially interesting in view of the expression of MICALs in the injured CNS. Following rat spinal cord lesions, *MICAL-1* and -3 signals are significantly increased in oligodendrocytes immediately adjacent to the site of injury. Furthermore, expression of all three MICALs is induced in meningeal fibroblasts that occupy the lesion core.[22] These observations suggest that MICALs may contribute to the regulation of post-injury responses such as neural scar formation.[30] Protein distribution patterns of vertebrate MICALs in the nervous system are currently unknown.

MICALs in Semaphorin Signaling

Invertebrate Signaling

Drosophila PlexA is a receptor for the transmembrane semaphorin Sema-1a, and PlexA-Sema-1a interactions are required for the generation of embryonic neuromuscular connectivity and for the establishment of longitudinal CNS axon tracts.[31,32] A yeast two-hybrid screen for PlexA-interacting proteins identified D-MICAL as a novel component of neuronal Sema-1a signaling.[19] Expression, biochemical, and genetic interaction experiments show a direct interaction between D-MICAL and PlexA which is required for the effects of Sema-1a on motor axon pathfinding.[19] How do these observations fit into our current view of Sema-1a/PlexA signaling? Similar to what has been found in vertebrate species, invertebrate semaphorin receptors are likely to be composed of multiple subunits. In addition to PlexA, the Sema-1a receptor complex is thought to contain at least one more component, off-track (Otk) (Fig. 2).[33] Otk, a putative receptor tyrosine kinase, associates with PlexA in vitro and modulates Sema-1a/PlexA signaling events through unidentified mechanisms. Another transmembrane protein that may function as a receptor component with PlexA is the receptor-type guanylyl cyclase Gyc76C (Fig. 2).[34] Genetic data reveal a neuronal requirement for the Gyc76C catalytic cyclase domain in Sema-1a repulsion. Although it remains to be determined whether or not Gyc76C associates directly with PlexA, Gyc76C may provide an in vivo link between semaphorin and cGMP signaling pathways previously characterized in vitro.[35] Similar to D-MICAL, the last four amino acids of Gyc76C fit the consensus for a PDZ (PSD-95, Discs-large, zona ocludens-1) domain-binding motif. This raises the intriguing possibility that, as has been observed for other signaling networks, PDZ domain containing scaffolding proteins may serve an important role in recruiting and assembling the Sema-1a signaling complex. Vertebrate semaphorin receptor subunits such as neuropilin-1 have been shown to regulate the interaction between plexinA receptors and their downstream effectors (see, for example ref. 36). It will be of great interest to determine whether Sema-1a coreceptor proteins such as Otk, or perhaps even Gyc76C, serve similar functions and thereby regulate the function of PlexA interacting proteins such as D-MICAL.

The intracellular signaling pathways downstream of the PlexA receptor complex are poorly defined. The A-kinase anchoring protein (AKAP) nervy and the small GTPase Rac are the only other cytosolic proteins, in addition to D-MICAL, that have thus far been implicated in Sema-1a/PlexA signaling.[37-39] Rac does not bind directly to PlexA, but increasing the level of *plexA* in vivo enhances the neuronal phenotype produced by the expression of a dominant-negative *Rac* mutant.[37,40] Nervy binds PlexA and couples this Sema-1a receptor to (cAMP-dependent) type II protein kinase A (PKA RII), providing the potential for spatiotemporal specific phosphorylation of target proteins under the control of local changes in cAMP (Fig. 2).[41] Interestingly, D-MICAL has several consensus PKA phosphorylation sites and, like nervy, binds to the PlexA cytoplasmic region. Although the exact D-MICAL and nervy binding sites on PlexA are unknown, genetic evidence argues against the idea that D-MICAL and nervy compete for PlexA binding, as has been suggested for certain RhoGTPases and vertebrate plexins.[15,19] Whether or not nervy is involved in regulating the activity of D-MICAL through its ability to localize cAMP-dependent PKAs remains to be determined. Vice versa, it is unknown whether D-MICAL influences the activity of Sema-1a receptor or signaling proteins such as Otk or nervy. However, the observation that site-directed mutagenesis of the D-MICAL monooxygenase region blocks Sema-1a-mediated axon repulsion supports the intriguing hypothesis that

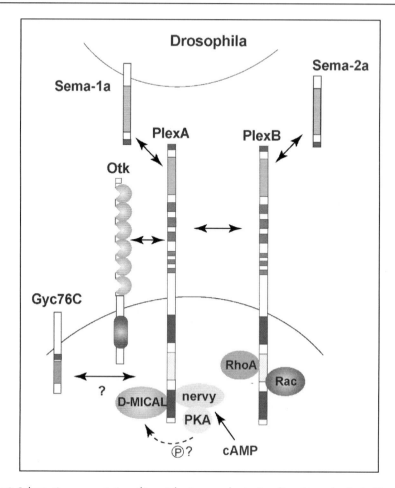

Figure 2. Schematic representation of invertebrate semaphorin signaling. Semaphorin-1a (Sema-1a) binds PlexA on the growth cone surface to influence axon pathfinding decisions in the *Drosophila* PNS and CNS. The receptor tyrosine kinase off-track (Otk) is a component of the PlexA receptor complex and is required for Sema-1a-mediated axon repulsion. The receptor-type guanylyl cyclase Gyc76C also functions with Sema-1a and PlexA and may be part of the PlexA receptor complex. The intracellular signaling cascades downstream of Sema-1a and PlexA are poorly understood but include at least two PlexA-binding proteins, D-MICAL and nervy. The function and mechanism-of-action of these signaling cues are largely unknown but it has been postulated that nervy, which is an A-kinase anchoring protein, may function to control the activity of D-MICAL through cAMP-dependent phosphorylation events. The other plexin receptor found in *Drosophila*, in addition to PlexA, is PlexB. PlexB is a functional receptor for Sema-2a and both Rac and RhoA have been reported to bind to PlexB. PlexA and PlexB can form heteromultimeric receptor complexes. This biochemical association allows for cooperative guidance effects and may enable PlexA access to PlexB-dependent signaling pathways (e.g., involving Rac), and vice versa (e.g., involving D-MICAL).

D-MICAL-mediated redox reactions function to regulate the activity of the Sema-1a/PlexA signaling complex.[19]

In addition to *plexA*, the *Drosophila* genome includes another plexin, *plexB*. PlexB is robustly expressed in the embryonic nervous system, is a functional receptor for Sema-2a, and is required for motor and CNS axon pathfinding.[42] Interestingly, PlexA and PlexB serve both distinct and shared neuronal guidance functions. For example, defects in the intersegmental nerve (ISN)b are strikingly similar in *plexA*- and *plexB*-deficient flies, while defects in longitudinally projecting CNS projections are clearly distinct.[31,42] The cooperative actions of PlexA and PlexB in patterning certain neuronal tracts can be explained by their ability to form heteromultimeric receptor complexes in vivo (Fig. 2).[42] Genetic and biochemical experiments suggest that this association enables PlexA access to signaling molecules that only bind to PlexB, and vice versa. For example, although D-MICAL does not bind directly to PlexB, a robust genetic interaction was observed between *D-MICAL* and *plexB* for ISNb defects.[42] Similarly, *Rac* genetically interacts with both *plexA* and *plexB* but fails to bind to PlexA (Fig. 2).[37,40] Expression analyses reveal widespread and overlapping *plexA* and *plexB* expression in the embryonic nervous system.[31,42] It is remains to be determined whether PlexA and PlexB associate simply as a result of coexpression in a subset of neurons or because of as yet unidentified molecular mechanisms. Genetic analyses further reveal that neuronal expression of *plexA* in a *plexB* mutant background reduces the severity of ISNb defects. In reciprocal experiments, PlexB cannot replace PlexA function.[42] These observations coupled with the genetic interactions between *plexB* and *D-MICAL* suggest that PlexA may be able to substitute for PlexB through its ability to recruit D-MICAL, while the inability of PlexB to substitute for PlexA may stem from its inability to directly recruit D-MICAL.

Vertebrate Signaling

Although the role of D-MICAL downstream of PlexA and PlexB has been firmly established, the neuronal function(s) of MICALs in vertebrate species remain largely unknown. However, current evidence hints at vertebrate MICALs contributing to Sema3/neuropilin/ plexinA signaling (Fig. 3). First, *MICAL, neuropilin,* and *plexinA* expression patterns overlap in several different regions of the nervous system.[22,43-45] Second, the COOH-terminal portion of human MICAL-1 interacts with the C2 cytoplasmic region of human plexinA3, while mouse MICAL-2 interacts with mouse plexinA4.[19] Third, pharmacological inhibitors with the ability to block flavoprotein monooxygenases similar to MICALs[46,47] neutralize axon repulsion induced by Sema3A and Sema3F, but not by Sema6A or additional unrelated repulsive guidance cues (Figs. 1B-E).[19,22] Overall, these experiments hint at the intriguing possibility that, similar to the invertebrate situation, vertebrate MICALs function in plexinA signaling. However, many questions remain about the putative role of MICALs in Sema3 repulsion. For example, do endogenous MICALs and plexinAs interact in neurons; are MICALs required for the Sema3 responsiveness of vertebrate neurons in vitro and in vivo? Future studies will undoubtedly address these and other issues.

Several distinct classes of cytosolic signaling proteins have been implicated in repulsive Sema3 signaling including CRMPs, RhoGTPases, and multiple intracellular protein kinases.[11,12,14] Unfortunately, our understanding of the linear and network relationships among these different cues is still rudimentary. Given their large multidomain structure, which allows for a plethora of protein-protein associations, MICALs are excellent candidates for recruiting and assembling critical components of the Sema3 signaling complex. Members of at least two classes of Sema3 signaling cues, the CRMPs and RanBPM, associate with MICALs (Fig. 3).[16,48-50] Whether any of the other cytosolic cues implicated in Sema3/plexinA signaling bind MICALs remains to be determined. However, the association of MICALs with several proteins at present thought not to be involved in signaling events involving plexinAs, CRMPs and RanBPM suggests that MICALs may not only function to transduce Sema3 signals but could also link Sema3 signaling to other signaling cascades, including those regulating cell adhesion or vesicle trafficking (see below).

A MICAL Connection to the Cytoskeleton?

In addition to an NH_2-terminal flavin adenine dinucleotide (FAD)-binding monooxygenase domain, MICALs contain a calponin homology (CH) domain, a LIM domain, proline-rich stretches and COOH-terminal coiled-coil motifs (Fig. 1A). Furthermore, available sequence

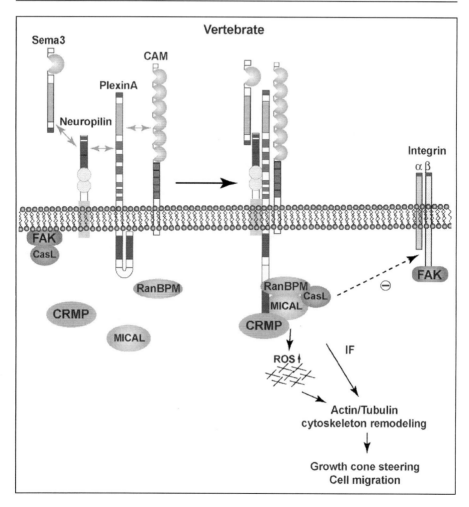

Figure 3. A potential role for MICAL proteins in vertebrate Sema3 signaling. Upon binding of Sema3s to neuropilins, a neuropilin/plexinA/CAM receptor complex is assembled and downstream signaling pathways are activated. Several lines of evidence suggest a role for MICALs downstream of Sema3s and plexinAs. Similar to the invertebrate situation, MICALs may bind the plexinA cytoplasmic domain to mediate repulsive Sema3 signals. At least two other classes of proteins involved in Sema3 signaling, i.e., CRMPs and RanBPM, have been reported to bind MICALs. In addition, MICALs bind CasL and may sequester this protein away from integrin signaling complexes in response to Sema3 binding thereby inhibiting cell adhesion. This figure is simplified to only contain MICALs and their Sema3 interacting partners. For a more comprehensive overview of Sema3 signaling see references within the text. CAM, cell adhesion molecule; CRMP, collapsin response mediator protein; FAK, focal adhesion kinase; ROS, reactive oxygen species. IF, intermediate filaments; RanBPM, Ran-small GTPase binding protein.

data suggest that MICAL-1, but not MICAL-2 and -3, contains a highly charged polyglutamic acid-rich stretch, and D-MICAL, but not mouse or human MICALs, contains a COOH-terminal PDZ domain binding motif. The founding member of the MICAL family, MICAL-1, was originally identified as a binding partner for CasL, a member of the p130^{Cas} (Cas) family.[18] MICAL-1 associates with the SH3 domain of CasL through a proline-rich

Pro-Pro-Lys-Pro-Pro (PPKPP) sequence near its COOH terminus and colocalizes with CasL in the perinuclear region of HeLa cells (Fig. 1A).[18] CasL acts as a docking protein for signaling molecules critical for β1-integrin-mediated formation of focal adhesions, sites where the actin cytoskeleton is attached to the extracellular matrix.[51,52] Interestingly, binding of semaphorins to plexins leads to a rapid disassembly of integrin-dependent focal adhesions, resulting in actin depolymerization and ultimately cell contraction.[53] Similarly, Sema3A inhibits integrin-mediated adhesion of endothelial cells to the extracellular matrix.[54] Together, these observations support the view that plexin-mediated inhibition of substrate adhesion is required for in vitro collapse responses. But how might plexins modulate integrin function? Several recent studies shed light on molecular mechanisms that may link plexin- and integrin-dependent signaling cascades. The central role of CasL in β1-integrin signaling coupled with its ability to bind MICAL-1, invite the speculation that MICALs may function to sequester CasL following plexin activation, thereby hampering integrin signaling and locally decreasing cell adhesion (Fig. 3). Other proposed, but not mutually exclusive, routes of plexin-induced inhibition of integrin signaling include the reduction of R-Ras and/or PIPKIγ661 kinase activity. Sema3 ligand binding triggers the dissociation of the RacGEF FARP2 from the plexinA1 receptor and as a result Rac1 is activated. Activated Rac1 facilitates binding of Rnd1 to plexinA1 which activates the plexinA1 GAP domain and stimulates GTPase activity for R-Ras. The resulting decrease in active R-Ras contributes to the suppression of integrin-mediated adhesion. After dissociation from plexinA1, FARP2 associates with and inhibits PIPKIγ661 kinase activity resulting in a further decrease in adhesion. Similar mechanisms have been proposed to underlie Sema4D-plexinB1 signaling events.[15,36,55,56] The disassembly of adhesive complexes as a result of plexin receptor activation is confirmed by the observation of decreased phosphorylation of focal adhesion components, such as focal adhesion kinase (FAK), following stimulation of NIH-3T3 cells expressing plexinB1 with Sema4D.[53] Like MICAL, FAK binds to CasL at its NH$_2$-terminal SH3 domain.[18,57] Thus, MICAL may not only sequester CasL but also block FAK-CasL associations and, as a consequence, modulate FAK phosphorylation and function (Fig. 3). Similarly, MICALs may interfere with the binding and function of other proteins that interact with the CasL SH3 domain including the protein tyrosine phosphatases (PTP)-1B and PTP-PEST, and the guanine nucleotide exchange factor C3G.[18]

Mammalian cells contain three types of cytoskeletal filaments: actin-containing filaments, tubulin-containing microtubules, and intermediate filaments (IFs). Emerging evidence suggests that MICALs may associate with or regulate the dynamics of several of these cytoskeletal components. The COOH-terminal region of MICAL-1 associates with the intermediate filament vimentin (Fig. 1A; amino acids 769-1067 H-MICAL-1), and MICAL-1 and vimentin expression patterns overlap in nonneuronal cells.[18] Although the biological significance of MICAL-vimentin interactions remains unknown, MICAL-1 could act as a novel IF-associated protein involved in maintaining cytoskeletal integrity, as has been shown for other IF-interacting protein such as MAP2 and plectin.[58,59]

Another MICAL-interactor is the small GTPase rab1. Rab1 binds MICAL-1, -2 and -3 at their COOH-terminus in a nucleotide-dependent manner[21,60] (Fig. 1A; aa 1028-1067 of H-MICAL-1). Rab1 has been shown to play a major role in the secretory pathway by targeting vesicles to their target destinations through interactions with the microtubule and/or actin cytoskeleton.[61,62] Interestingly, the spatial distribution of EGFP-tagged MICAL-1 and -3 in Hela cells resembles that of cytoskeletal components such as tubulin and actin, and treatment with nocodazol, a microtubule depolymerizing agent, results in a loss of this MICAL network-like staining.[60] Similarly, MICAL-L2 colocalizes with the actin cytoskeleton in MTD1A and NIH-3T3 cells, and treatment with cytochalasin D, an actin depolymerizing agent, disrupts MICAL-L2 distribution patterns.[20] This, together with the ability of MICAL-L2 to bind another member of the rab family (rab13)[20] suggests that MICALs may function in linking rab GTPases to the cytoskeleton.

Although there are no additional MICAL-interacting molecules known to date, there are several well-conserved domains within the MICAL coding sequence that could serve to link

MICALs to the cytoskeleton. Although the function and mechanism-of-action of the NH_2-terminal MICAL flavoprotein monooxygenase domain is unknown, redox signaling modification of amino acids within cytoskeletal or signaling proteins has been shown to modulate their function.[63,64,79] In addition, oxidation of actin leads to disassembly of actin filaments, collapse of actin networks, reduced interactions between actin and actin crosslinking proteins, and decreased actin polymerization.[65-67] Calponin homology (CH) domains are found in many other adaptor proteins and have been implicated in both actin- and microtubule-binding.[68] Finally, LIM domains, conserved double-zinc finger motifs, are found in various cytoskeleton regulatory proteins.[69,70] Future studies focusing on the identification and characterization of proteins that bind the various MICAL domains will be essential for providing novel insight into the biological function(s) of MICALs, including their putative role as cytoskeletal regulators.

MICALs: Redox Regulators of Axon Guidance Events?

Sequence, structural and biophysical analyses show that the highly conserved NH_2-terminal portion of murine MICAL-1 has the architecture and characteristics of a flavoenzyme of the monooxygenase family (Fig. 1F).[19,71,72] Flavoprotein monooxygenases are oxidoreductases that use FAD to catalyze the insertion of one atom of molecular oxygen into their substrates. In addition, in some contexts these enzymes can act as oxidases and generate reactive oxygen species (ROS).[73,74] Two lines of experimental evidence suggest that the MICAL monooxygenase domain is required for mediating repulsive semaphorin function. Site-directed mutagenesis of functional residues in the D-MICAL monooxygenase region attenuates repulsive Sema-1a signaling in vivo and flavoprotein monooxygenase inhibitors neutralize Sema3-induced axon repulsion in vitro.[19,22] These observations hint at a novel role for redox reactions in repulsive semaphorin signaling.

The topology of the MICAL-1 flavoenzyme most closely resembles that of p-hydroxybenzoate hydroxylase (PHBH), a nicotinamide adenine dinucleotide phosphate (NADPH)-dependent flavoprotein monooxygenase.[75] Comparison of the MICAL-1 structure before and after addition of NADPH reveals that, similar to PHBH, the MICAL-1 flavin ring can switch between two discrete positions ('in' and 'out'). In contrast to other monooxygenases, this conformational change is coupled with the opening of a channel to the active site, suggestive of a protein substrate (Fig. 4).[72] This is especially interesting since the substrates of hydroxylases have classically been thought to be small molecules[76] (e.g., p-hydroxybenzoate, steroids and amino acids). The monooxygenase domain of all MICALs contains an extensive patch of basic potential, suggestive of a binding surface for a positively charged protein substrate.[72] It is noteworthy in this regard that both actin and several actin-related proteins are highly acidic.[77] Alternatively, other MICAL domains (e.g., calponin homology and LIM domains) may function to bind flavoenzyme substrates and present them to the monooxygenase region (Fig. 4). Determination of the structure of other MICAL domains in combination with the monooxygenase domain will help to validate these different scenarios.

In addition to directly modifying potential substrates through redox reactions, MICALs may influence intracellular signaling events through the generation of ROS (Fig. 4). It has long been thought that ROS such as O_2^-, H_2O_2, OH• and NO were predominantly damaging to the cell because of their reactive nature and ability to alter the integrity of macromolecules such as DNA, lipids and proteins. Recently however, it has become clear that the restricted generation of ROS is physiologically essential in several signal transduction pathways. Regulated oxidation and reduction (redox) reactions can modify transcription factors or enzymes, including GTPases, in order to regulate basic cellular functions such as growth, differentiation and adaptation to external stimuli.[63,78,79] Many of these responses are elicited by cytokines and growth factors resulting in transient bursts of ROS.[80-82] In the presence of NADPH, MICAL-1 can reduce molecular oxygen to H_2O_2, a signaling molecule involved in a multitude of biological processes including protein phosphorylation.[67,83,84] It is now also widely accepted that ROS can alter the integrity of the actin cytoskeleton.[85,86] One example of ROS regulation of

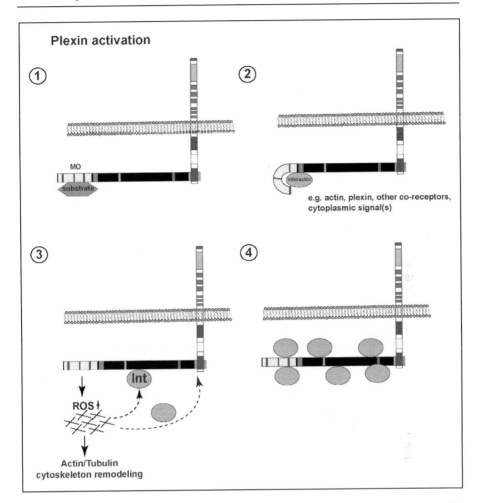

Figure 4. How MICALs may mediate plexin signaling. 1) The flavoprotein monooxygenase (MO) domain of MICAL may modulate the activity of a substrate to influence downstream signaling events. 2, 3) Cytoskeletal components, semaphorin (co)receptors or other cytosolic and transmembrane proteins bound to or in the vicinity of the MICAL protein may be modified by the MO domain directly (2) or through the production of reactive oxygen species (ROS) (3). 4) MICAL acts as a protein scaffold and forms a multi-component signaling complex that mediates semaphorin-induced axon steering events. Int: interactor.

the actin cytoskeleton is ROS-directed depolymerization of F-actin under the control of the small GTPase Rac1.[87,88] Rac1-mediated activation of NAD(P)H oxidase generates ROS which can downlegulate Rho activity leading to the formation of membrane ruffles and cell movement.[89] Whether or how MICAL flavoenzyme-mediated redox signaling is exactly involved in the regulation of the neuronal cytoskeleton remains to be determined.

Concluding Remarks

Genetic and biochemical evidence supports a prominent role for D-MICAL in repulsive semaphorin signaling. Site-directed mutagenesis and pharmacological inhibitor studies furthermore indicate that the evolutionary conserved MICAL flavoprotein monooxygenase domain is likely essential for mediating repulsive semaphorin signaling in vivo, supporting a novel and unexpected role for redox signaling in axon guidance. In addition, their large multidomain structure suggests that MICALs may serve as protein scaffolds to recruit, assemble and/or activate (part of) the semaphorin signaling network. However, as outlined here, many questions remain about MICAL function and mechanism-of-action. For example, do MICALs directly bind to cytoskeletal components, and is MICAL-induced redox signaling involved in regulating cytoskeletal dynamics? What is the substrate of the MICAL monooxygenase domain, and how is the activity of this flavoenzyme regulated? Do vertebrate MICALs, similar to D-MICAL, function in semaphorin/plexinA signaling? Semaphorins are implicated in the onset and/or progression of human brain disease and have been proposed to act as molecular inhibitors of regenerating axons in the injured CNS. Given the prominent expression of MICALs in the intact and injured adult nervous system[22] it will be of great interest to examine whether MICALs can serve as therapeutic targets for counteracting or controlling semaphorin-mediated effects in the injured or diseased brain.

Acknowledgements

We would like to thank Peter Burbach, Alex Kolodkin and Jon Terman for critical reading of the manuscript and Yvonne Jones and Christian Siebold for providing us with Figure 1F. Work in the laboratory of the authors is supported by grants from the Netherlands Organization for Scientific Research (to SMK and RJP), the Dutch Brain Foundation (to SMK and RJP), the Human Frontier Science Program Organization, the Genomics Center Utrecht (to RJP). RJP is a NARSAD Henry and William Test Investigator.

References

1. Tessier-Lavigne M, Goodman CS. The molecular biology of axon guidance. Science 1996; 274(5290):1123-33.
2. Dickson BJ. Molecular mechanisms of axon guidance. Science 2002; 298(5600):1959-64.
3. Huber AB, Kolodkin AL, Ginty DD et al. Signaling at the growth cone: Ligand-receptor complexes and the control of axon growth and guidance. Annu Rev Neurosci 2003; 26:509-63.
4. Raper JA. Semaphorins and their receptors in vertebrates and invertebrates. Curr Opin Neurobiol 2000; 10:88-94.
5. Fiore R, Puschel AW. The function of semaphorins during nervous system development. Front Biosci 2003; 8:s484-99.
6. Casazza A, Fazzari P, Tamagnone L. Semaphorin signals in cell adhesion and cell migration: Functional role and molecular mechanisms. THIS BOOK 2006.
7. Neufeld G, Lange T, Varshavsky A et al. Semaphorin signaling in vascular and tumor biology. THIS BOOK 2006.
8. Potiron V, Nassare P, Roche J et al. Semaphorin signaling in the immune system. THIS BOOK 2006.
9. Toyofuku T, Kikutani H. Semaphorin signaling during cardiac development. THIS BOOK 2006.
10. Bechara A, Falk J, Moret F et al. Modulation of Semaphorin signaling by Ig super family cell adhesion molecules. THIS BOOK 2006.
11. Kruger RP, Aurandt J, Guan KL. Semaphorins command cells to move. Nat Rev Mol Cell Biol 2005; 6(10):789-800.
12. Pasterkamp RJ, Kolodkin AL. Semaphorin junction: Making tracks toward neural connectivity. Curr Opin Neurobiol 2003; 13:79-89.
13. Ahmed A, Eickholt B. Intracellular kinases in Semaphorin signaling. THIS BOOK 2006.
14. Castellani V, Rougon P. Control of semaphorin signaling. Curr Opin Neurobiol 2002; 12:532-541.
15. Puschel AW. GTPases in semaphorin signaling. THIS BOOK 2006.
16. Schmidt EF, Strittmatter SM. The CRMP family of proteins and their role in Sema3A signaling. THIS BOOK 2006.
17. Shim S, Ming Gl. Signaling of secreted semaphorins in growth cone steering. THIS BOOK 2006.
18. Suzuki T, Nakamoto T, Ogawa S et al. MICAL, a novel CasL interacting molecule, associates with vimentin. J Biol Chem 2002; 277:14933-14941.
19. Terman JR, Mao T, Pasterkamp RJ et al. MICALs, a family of conserved flavoprotein oxidoreductases, function in plexin-mediated axonal repulsion. Cell 2002; 109:887-900.
20. Terai T, Nishimura N, Kanda I et al. JRAB/MICAL-L2 is a junctional Rab13-binding protein mediating the endocytic recycling of occludin. Mol Biol Cell 2006; 17(5):2465-2475.
21. Weide T, Teuber J, Bayer M et al. MICAL-1 isoforms, novel rab1 interacting proteins. Biochem Biophys Res Commun 2003; 306:79-86.

22. Pasterkamp RJ, Dai H, Terman JR et al. MICAL flavoprotein monooxygenases: Expression during neural development and following spinal cord injuries in the rat. Mol Cell. Neurosci 2006; 31:52-69.
23. Cohen RI, Rottkamp DM, Maric D et al. A role for semaphorins and neuropilins in oligodendrocyte guidance. J Neurochem 2003; 85:1262-78.
24. Goldberg JL, Vargas ME, Wang JT et al. An oligodendrocyte lineage-specific semaphorin, Sema5A, inhibits axon growth by retinal ganglion cells. J Neurosci 2004; 24(21):4989-99.
25. Moreau-Fauvarque C, Kumanogoh A, Camand E et al. inhibitor of axonal growth, is expressed on oligodendrocytes and upregulated after CNS lesion. J Neurosci 2003; 8;23(27):9229-39.
26. Ricard D, Stankoff B, Bagnard D et al. Differential expression of collapsin response mediator proteins (CRMP/ULIP) In subsets of oligodendrocytes in the postnatal rodent brain. Mol Cell Neurosci 2000; 16(4):324-37.
27. Ricard D, Rogemond V, Charrier E et al. Isolation and expression pattern of human Unc-33-like phosphoprotein 6/collapsin response mediator protein 5 (Ulip6/CRMP5): Coexistence with Ulip2/CRMP2 in Sema3a- sensitive oligodendrocytes. J Neurosci 2001; 21:7203-7214.
28. Spassky N, de Castro F, Le Bras B et al. Directional guidance of oligodendroglial migration by class 3 semaphorins and netrin-1. J Neurosci 2002; 22:5992-6004.
29. Giraudon P, Vincent P, Vuaillat C et al. Semaphorin CD100 from activated T lymphocytes induces process extension collapse in oligodendrocytes and death of immature neural cells. J Immunol 2004; 172(2):1246-55.
30. Pasterkamp RJ, Verhaagen J. Semaphorins in axon regeneration: Developmental guidance molecules gone wrong? Royal Soc Phil Trans B 2006, (in press).
31. Winberg ML, Noordermeer JN, Tamagnone L et al. Plexin A is a neuronal semaphorin receptor that controls axon guidance. Cell 1998; 95:903-16.
32. Yu HH, Araj HH, Ralls SA et al. The transmembrane Semaphorin Sema I is required in Drosophila for embryonic motor and CNS axon guidance. Neuron 1998; 20(2):207-20.
33. Winberg ML, Tamagnone L, Bai J et al. The transmembrane protein Off-track associates with Plexins and functions downstream of Semaphorin signaling during axon guidance. Neuron 2001; 32:53-62.
34. Ayoob JC, Yu HH, Terman JR et al. The Drosophila receptor guanylyl cyclase Gyc76C is required for semaphorin-1a-plexin A-mediated axonal repulsion. J Neurosci 2004; 24(30):6639-49.
35. Song H, Ming G, He Z et al. Conversion of neuronal growth cone responses from repulsion to attraction by cyclic nucleotides. Science 1998; 281(5382):1515-8.
36. Toyofuku T, Yoshida J, Sugimoto T et al. FARP2 triggers signals for Sema3A-mediated axonal repulsion. Nat Neurosci 2005; 8(12):1712-9.
37. Hu H, Marton TF, Goodman CS. Plexin B mediates axon guidance in Drosophila by simultaneously inhibiting active Rac and enhancing RhoA signaling. Neuron 2001; 32:39-51.
38. Bashaw GJ. Semaphorin signaling unplugged: A Nervy AKAP cAMP(s) out on Plexin. Neuron 2004; 42:363-366.
39. Terman JR, Kolodkin AL. Nervy links protein kinase a to plexin-mediated semaphorin repulsion. Science 2004; 303(5661):1204-7.
40. Driessens MHE, Hu H, Nobes CD et al. Plexin B semaphorin receptors interact directly with active Rac and regulate the actin cytoskeleton by activating Rho. Curr Biol 2001; 11:339-344.
41. Feliciello A, Gottesman ME, Avvedimento EV. The biological functions of A-kinase anchor proteins. J Mol Biol 2001; 308(2):99-114.
42. Ayoob JC, Terman JR, Kolodkin AL. Drosophila Plexin B is a Sema-2a receptor required for axon guidance. Development 2006; 133, (in press).
43. Kawakami A, Kitsukawa T, Takagi S et al. Developmentally regulated expression of a cell surface protein, neuropilin, in the mouse nervous system. J Neurobiol 1996; 29(1):1-17.
44. Murakami Y, Suto F, Shimizu M et al. Differential expression of plexin-A subfamily members in the mouse nervous system. Dev Dyn 2001; 220(3):246-58.
45. Suto F, Murakami Y, Nakamura F et al. Identification and characterization of a novel mouse plexin, plexin-A4. Mech Dev 2003; 120(3):385-96.
46. Abe I, Kashiwagi K, Noguchi H. Antioxidative galloyl esters as enzyme inhibitors of p-hydroxybenzoate hydroxylase. FEBS Lett 2000; 483:131-134.
47. Abe I, Seki T, Noguchi H. Potent and selective inhibition of squalene epoxidase by synthetic galloyl esters. Biochem Biophys Res Commun 2000; 270:137-140.
48. Schmidt EF, Togashi H, Strittmatter SM. Characterization of a multi-molecular signaling complex that mediates Sema3A signaling. Soc Neurosci Abstracts 2004.
49. Deo RC, Schmidt EF, Elhabazi A et al. Structural bases for CRMP function in plexin-dependent semaphorin3A signaling. EMBO J 2004; 23(1):9-22.
50. Togashi H, Schmidt EF, Strittmatter SM. RanBPM contributes to Semaphorin3A signaling through Plexin-A receptors. J Neurosci 2006; 26(18):4961-4969.

51. O'Neill GM, Fashena SJ, Golemis EA. Integrin signaling: A new Cas(t) of characters enters the stage. Trends Cell Biol 2000; 10:111-119.
52. Yi J, Kloeker S, Jensen CC et al. Members of the Zyxin family of LIM proteins interact with members of the p130Cas family of signal transducers. J Biol Chem 2002; 277:9580-9589.
53. Barberis D, Artigiani S, Casazza A et al. Plexin signaling hampers integrin-based adhesion, leading to Rho-kinase independent cell rounding, and inhibiting lamellipodia extension and cell motility. FASEB J 2004; 18:592-604.
54. Serini G, Valdembri D, Zanivan S et al. Class 3 semaphorins control vascular morphogenesis by inhibiting integrin function. Nature 2003; 424(6947):391-7.
55. Oinuma I, Katoh H, Negishi M. Molecular dissection of the semaphorin 4D receptor plexin-B1-stimulated R-Ras GTPase-activating protein activity and neurite remodeling in hippocampal neurons. J Neurosci 2004; 24(50):11473-80.
56. Negishi M, Oinuma I, Katoh H. R-ras as a key player for signaling pathway of plexins. Mol Neurobiol 2005; 32(3):217-22.
57. Kamiguchi K, Tachibana K, Iwata S et al. Cas-L is required for beta 1 integrin-mediated costimulation in human Tcells. J Immunol 1999; 163:563-568.
58. Foisner R, Wiche G. Intermediate filament-associated proteins. Curr Opin Cell Biol 1991; 3(1):75-81.
59. Hirokawa N, Hisanaga S, Shiomura Y. MAP2 is a component of crossbridges between microtubules and neurofilaments in the neuronal cytoskeleton: Quick-freeze, deep-etch immunoelectron microscopy and reconstitution studies. J Neurosci 1988; 8(8):2769-79.
60. Fischer J, Weide T, Barnekow A. The MICAL proteins and rab1: A possible link to the cytoskeleton? Biochem Biophys Res Commun 2005; 328:415-423.
61. Seabra MC, Coudrier E. Rab GTPases and myosin motors in organelle motility. Traffic 2004; 5(6):393-9.
62. Murshid A, Presley JF. ER-to-Golgi transport and cytoskeletal interactions in animal cells. Cell Mol Life Sci 2004; 61(2):133-45.
63. Finkel T. Oxygen radicals and signaling. Curr Opin Cell Biol 1998; 10:248-253.
64. Kamata H, Honda S, Maeda S et al. Reactive oxygen species promote TNFalpha-induced death and sustained JNK activation by inhibiting MAP kinase phosphatases. Cell 2005; 120(5):649-61.
65. Dalle-Donne I, Rossi R, Milzani A et al. The actin cytoskeleton response to oxidants: From small heat shock protein phosphorylation to changes in the redox state of actin itself. Free Radic Biol Med 2001; 31(12):1624-32.
66. Dalle-Donne I, Rossi R, Giustarini D et al. Actin carbonylation: From a simple marker of protein oxidation to relevant signs of severe functional impairment. Free Radic Biol Med 2001; 31:1075-83.
67. Milzani A, Dalle-Donne I, Colombo R. Prolonged oxidative stress on actin. Arch Biochem Biophys 1997; 339(2):267-74.
68. Gimona M, Djinovic-Carugo K, Kranewitter WJ et al. Functional plasticity of CH domains. FEBS Lett 2002; 513:98-106.
69. Bach I. The LIM domain: Regulation by association. Mech Dev 2000; 91(1-2):5-
70. Kadrmas JL, Beckerle MC. The LIM domain: From the cytoskeleton to the nucleus. Nat Rev Mol Cell Biol 2004; 5:920-31.
71. Nadella M, Bianchet MA, Gabelli SB et al. Structure and activity of the axon guidance protein MICAL. Proc Natl Acad Sci USA 2005; 102(46):16830-5.
72. Siebold C, Berrow N, Walter TS et al. High-resolution structure of the catalytic region of MICAL, a multi-domain flavoenzyme-signaling molecule. Proc Natl Acad Sci USA 2005; 102:16836-16841.
73. Massey V. Activation of molecular oxygen by flavins and flavoproteins. J Biol Chem 1994; 269(36):22459-62.
74. Massey V. Introduction: Flavoprotein structure and mechanism. FASEB J 1995; 9(7):473-5.
75. Wierenga RK, de Jong RJ, Kalk KH et al. Crystal structure of p-hydroxybenzoate hydroxylase. J Mol Biol 1979; 131(1):55-73.
76. Ghisla S, Massey V. Mechanisms of flavoprotein-catalyzed reactions. Eur J Biochem 1989; 181(1):1-17.
77. Otterbein LR, Graceffa P, Dominguez R. The crystal structure of uncomplexed actin in the ADP state. Science 2001; 293(5530):708-11.
78. Finkel T. Redox-dependent signal transduction. FEBS Lett 2000; 476:52-54.
79. Kamata H, Hirata H. Redox regulation of cellular signaling. Cell Signal 1999; 11:1-14.
80. Lo YYC, Cruz TF. Involvement of reactive oxygen species in cytokine and growth factor induction of c-fos expression in chondrocytes. J Biol Chem 1995; 270:11727-11730.
81. Sundaresan M, Yu ZX, Ferrans VJ et al. Requirement for generation of H_2O_2 for platelet derived growth factor signal transduction. Science 1995; 270:296-299.

82. Svegliati S, Cancello R, Sambo P et al. Platelet-derived growth factor and reactive oxygen species (ROS) regulate Ras protein levels in primary human fibroblasts via ERK1/2. Amplification of ROS and Ras in systemic sclerosis fibroblasts. J Biol Chem 2005; 280:36474-82.
83. Rhee SG. Redox signaling: Hydrogen peroxide as intracellular messenger. Exp Mol Med 1999; 31:53-59.
84. Rhee SG, Bae YS, Lee SR et al. Hydrogen peroxide: A key messenger that modulates protein phosphorylation through cysteine oxidation. Sci STKE 2000; 53:PE1.
85. Guay J, Lambert H, Gingras-Breton G et al. Regulation of actin filament dynamics by p38 map kinase-mediated phosphorylation of heat shock protein 27. J Cell Sci 1997; 110(Pt 3):357-68.
86. Ahn SG, Thiele DJ. Redox regulation of mammalian heat shock factor 1 is essential for Hsp gene activation and protection from stress. Genes Dev 2003; 17(4):516-28.
87. Kheradmand F, Werner E, Tremble P et al. Role of Rac1 and oxygen radicals in collagenase-1 expression induced by cell shape change. Science 1998; 280:898-902.
88. Moldovan L, Irani K, Moldovan NI et al. The actin cytoskeleton reorganization induced by Rac1 requires the production of superoxide. Antioxid Redox Signal 1999; 1:29-43.
89. Nimnual AS, Taylor LJ, Bar-Sagi D. Redox-dependent downregulation of Rho by Rac. Nat Cell Biol 2003; 5:236-241.

CHAPTER 5

Signaling of Secreted Semaphorins in Growth Cone Steering

Sangwoo Shim and Guo-li Ming*

Introduction

Secreted semaphorins [class 3 semaphorins (3A-3F) in vertebrates and class 2 in invertebrates] play essential roles in the establishment of neuronal circuitry by mediating axon steering and fasciculation during development of the nervous system.[1-3] Semaphorin 3A (Sema3A) was the first secreted form of semaphorins purified from adult chick brains and characterized as a chemorepellant with the ability to induce rapid collapse or repulsion of dorsal root ganglion (DRG) growth cones in vitro and to repel populations of neurons in vivo.[4-8] In the grasshopper, a graded distribution of Sema-2a has been shown to be essential in guiding the tibial (Ti1) pioneer neurons in the developing limb.[9] Like other families of guidance cues, such as netrins and ephrins, semaphorins function not only as repellents but also as attractants to neuronal growth cones, depending on the composition of receptors and signaling cascades presented in the cells. Sema3C, for example, can act as a chemoattractant to embryonic cortical axons.[10] Sema3B has recently been shown to attract and repel commissural axons in vitro and is critical for the positioning of anterior commissural projection.[11] In a slice overlay assay, Sema3A was shown to attract the dendritic growth cones of cortical neurons,[12] while repelling the axonal growth cones of the same neurons.[13] The molecular mechanisms in mediating and modulating growth cone steering responses to class 3 semaphorins have been best characterized within the semaphorin family both in vitro and in vivo, and are the main focus of this chapter. Interested readers can consult other chapters of the book and several other comprehensive reviews on semaphorins and their signaling.[3,14-16]

In Vitro Neuronal Growth Cone Steering Assays

Since in vitro assays have been indispensable in determining the function and molecular mechanisms of class 3 semaphorin signaling in growth cone steering, we will first briefly introduce and describe some of these in vitro assays. For all growth cone steering assays, a gradient of the semaphorin protein is produced, either by being released from source cells (natural semaphorin producing cells or cell lines transfected with semaphorin expression constructs), or from a micropipette loaded with purified recombinant protein.

Growth Cone Turning Assay[17,18]

This assay has been extensively used with *Xenopus* spinal neurons, retinal ganglion cells and mammalian neurons to determine signaling mechanisms of various diffusible guidance cues, including class 3 secreted semaphorins. A microscopic gradient is established by controlled

*Corresponding Author: Guo-li Ming—Institute for Cell Engineering, Departments of Neurology and Neuroscience, Johns Hopkins University School of Medicine, Baltimore, Maryland 21025, U.S.A. Email: gming1@bs.jhmi.edu

Semaphorins: Receptor and Intracellular Signaling Mechanisms, edited by R. Jeroen Pasterkamp.
©2007 Landes Bioscience and Springer Science+Business Media.

pulsatile ejection of solutions containing semaphorins from a micropipette and the growth cone responses within the gradient can be quantified by the degree of the turning angle with respect to the original direction of neurite extension. The in vitro gradient is very stable and reliable, thus allowing quantitative analysis of the steering decision of individual neuronal growth cones in response to a defined gradient of guidance cues. In addition, pharmacological and genetic manipulations as well as high resolution imaging of cellular events (e.g., Ca^{2+} imaging) are relatively easy to be achieved in this system.

Collagen Gel Repulsion Assay[19,20]

In this assay, neuronal tissue explants and aggregates of COS cells expressing class 3 secreted semaphorins are embedded in three-dimensional collagen gels for coculture of a few days. Axonal outgrowth is then visualized by fixation and immunostaining with antibodies to neurofilament. The chemotropic activity of the protein of interest is quantified by the axon outgrowth ratio P/D, where P is the extent of axonal outgrowth on the side proximal to the COS cell aggregate, and D is the extent of axonal outgrowth distal to the cell aggregate. A P/D ratio below one, therefore, indicates chemorepulsion.

Slice Overlay Assay[12,21]

In this assay, dissociated cells isolated from the developing nervous system are labeled and cultured over neural tissue slices from various developmental stages and regions. This system can potentially provide individual neuronal growth cones with an environment that better mimics the endogenous milieu. The axonal or dendritic orientation of dissociated cells in response to guidance cues provided by the slice is then analyzed by the morphology of neurons.

Receptor Complex in Mediating Growth Cone Turning Responses to Class 3 Semaphorins

The chemotropic effects of class 3 semaphorins on growth cone steering are mediated by a functional receptor complex comprised of neuropilins (Neuropilin-1 and Neuropilin-2) as the ligand-binding component[22-26] and Plexin-As (A1-A4) as the signal transducing component (Fig. 1).[27-29] The neuropilin/plexin receptor complex appears to have different binding specificity for class 3 semaphorins and exhibits mostly complementary and distinct temporal and spatial expression profiles in developing neurons of both the central (CNS) and the peripheral nervous systems (PNS), which may explain the neuronal subtype specificity of chemotropic effects of class 3 semaphorins.

The growth cones of cultured embryonic *Xenopus* spinal neurons or retinal ganglion cells exhibit robust repulsive turning responses to a gradient of Sema3A in a chemotropic growth cone turning assay.[18,30] Similar repulsive responses were observed in a collagen gel repulsion assay using mammalian sensory neurons.[31] Neuropilin-1 and Plexin-A1 mediate Sema3A-induced repulsion in these neurons, since application of a function-blocking antiserum against the extracellular domain of Neuropilin-1 or overexpression of a truncated form of Plexin-A1 lacking the highly conserved cytoplasmic domain completely abolished the growth cone repulsion.[18,29,31] In addition, dendritic growth cones also require Neuropilin-1 function for Sema3A-induced guidance responses. Experiments using function-blocking antibody in a slice overlay assay showed that Neuropilin-1 serves as a Sema3A receptor in mediating both the chemoattractive guidance of cortical apical dendrites towards the pial surface and the chemorepulsive guidance of cortical axons toward the white matter in response to a Sema3A gradient.[12,13]

Similar to Sema3A, other members of the class 3 semaphorin family (Sema3B, 3C, 3D, 3E, 3F) also exhibit chemotropic guidance activities in vitro but differ in their binding specificities for neuropilins. For example, Sema3B, 3C and 3F appear to preferentially bind to the Neuropilin-2 homodimer or Neuropilin-1/2 heterodimer, whereas Sema3A preferentially binds to Neuropilin-1 homodimers.[24-26] It has been shown that sympathetic axons coexpress

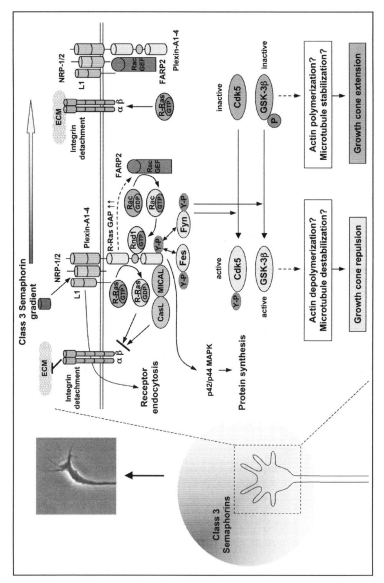

Figure 1. Components of class 3 secreted semaphorin signaling for growth cone steering. In neuronal growth cones, class 3 secreted semaphorins bind to a receptor complex with Plexin-As (A1-4) and Neuropilins (Neuropilin-1 or Neuropilin-2) to propagate a number of signaling pathways. Sema3A binding to Neuropilin-1 triggers the dissociation of FARP2 and activation of RacGEF, leading to the activation of plexin-A1 downstream signaling events, such as activation of R-Ras GAP of plexin-A1 and subsequent down regulation of R-Ras, which in turn limits integrin-mediated attachment. The activated semaphorin receptor complex also stimulates the activity of protein kinases Fes, Fyn, Cdk5 and GSK-3β and these activations may induce biased cytoskeletal reorganization, such as actin polymerization and microtubule assembly, thereby leading to neuronal growth cone repulsion.

Neuropilin-1 and Neuropilin-2, whereas sensory axons express only Neuropilin-1, therefore, sensory axons are sensitive to Sema3A, whereas sympathetic axons show responsiveness to all three Sema3s.[23,24] Interestingly, Sema3B and Sema3C were shown to block Sema3A binding to Neuropilin-1 and abolish the repellent actions of Sema3A on sensory neurons by competing the binding to Neuropilin-1.[25]

In addition to Plexin-A1, other Plexin-A family members are involved in transducing the signals of class 3 semaphorins, but with less specificity. Overexpression of a dominant negative form of Plexin-A2 in DRG sensory neurons renders their axons insensitive to Sema3A,[31] suggesting that Plexin-A2 also partially contributes to the receptor signaling of Sema3s. In a more recent study using the collagen gel repulsion assay, analysis of Plexin-A3, Plexin-A4 and Plexin-A3/-A4 double knockout mice showed that while Plexin-A3 and -A4 together mediate the responses to class 3 semaphorins in both sensory and sympathetic neurons, Sema3A repulsive signaling is mediated principally by Plexin-A4 via Neuropilin-1 and Sema3F repulsive signaling is mediated principally by Plexin-A3 via Neuropilin-2.[32,33]

The specificity of receptor binding and repulsive effects of different class 3 semaphorins seems to be conserved in both PNS and CNS. Differential expression patterns of class 3 semaphorins[34,35] and their receptors (neuropilins[24,35,36] and plexins[32,37]) in the developing hippocampus and afferent connections have also been implicated in the specificity of the chemorepulsive actions of different class 3 semaphorins.[32,34-36] Notably, hippocampal axons explanted from the embryonic dentate gyrus (DG), CA1 and CA3 regions express both Neuropilin-1 and Neuropilin-2 and are repelled by both Sema3A and Sema3F in a collagen gel repulsion assay.[34] Moreover, function-blocking antibodies against Neuropilin-1 block the repulsive effect of Sema3A but not Sema3F,[34] while hippocampal axons from Neuropilin-2 knockout mice (Nrp2-/-) lose their responsiveness to Sema3F but not Sema3A.[38] Analysis of Plexin-A3 knockout mice with collagen gel repulsion assays showed that Plexin-A3 also contributes to chemorepulsive effects of Sema3A and Sema3F on hippocampal axons.[32]

Intracellular Mediators for Class 3 Semaphorin-Induced Growth Cone Turning

While the intracellular pathways, from the receptor activation to changes of cytoskeleton proteins that result in growth cone steering in response to class 3 semaphorins, are still not fully understood, several molecules and mechanisms have been identified to be involved in the cytoplasmic signaling of this family of guidance cues over the past decade.

Protein Kinases

After the initial observation of tyrosine phosphorylation of plexins in vitro,[29] several cytoplasmic kinases have been implicated in semaphorin-plexin-mediated growth cone responses (Fig. 1). For example, tyrosine kinase Fes,[39] threonine-serine kinase cyclin-dependent kinase 5 (Cdk5)[40] and glycogen synthase kinase-3 (GSK-3)[41] have been shown to mediate sema3A-induced growth cone collapse. However, the functional roles of these molecules in growth cone steering have not been fully explored. A correlative study using Sema3A and fyn knockout mice also suggests a role of Fyn, a Src family nonreceptor tyrosine kinase, in mediating the guidance of apical dendrites of large pyramidal neurons to sema3A.[40] In cortical slices prepared from null mutants, some pyramidal neurons exhibit an atypical morphology of dendritic orientation in both fyn-/- and Sema3A-/- cortices using Golgi impregnation analysis. This study, however, did not provide direct evidence for the role of Fyn in dendritic guidance in response to Sema3A.

MICAL

The MICAL (molecule interacting with Cas ligand) family of cytosolic, multidomain, flavoprotein monooxygenases has recently been identified as a binding partner to plexins and is involved in semaphorin-plexin-mediated axon guidance.[42] Interestingly, the monooxygenase enzyme activity seems to be required for growth cone turning responses to Sema3A and 3F. In a collagen gel repulsion assay, a flavoprotein monooxygenase inhibitor neutralizes Sema3A- and Sema3F-mediated repulsion of both DRG and superior cervical ganglion (SCG) axons, respectively.[42,43] These results, together with a recent report that 12/15-lipoxygenase is required for Sema3A-mediated axonal collapse,[44] suggest an important role of oxidation on the regulation of inhibitory effects of class 3 semaphorins on neuronal growth cones.

Rho Family of GTPases

Direct growth cone turning is mediated by biased cytoskeletal reorganization localized within a growth cone. Small GTPases of the Rho family (which include Rac, Rho and Cdc42) provide an important link between semaphorin receptor signaling and cytoskeletal dynamics in neurons.[45] Although many of the Rho GTPases, such as Rnd, Rac and Rho, have been shown to mediate semaphorin signaling in other biological events, their roles in mediating growth cone turning has not been firmly established. Their regulators, including activator GEFs (guanine exchange factors) and inactivator GAPs (GTPase activating proteins), however, have been shown to be involved in Sema3A-induced growth cone turning. Recently, a FERM domain-containing Rac-GEF protein, FARP2, has been identified to serve as a physical link between Sema3A binding and Rac activation.[46,47] Sema3A-induced repulsion of axons of DRG neurons was completely abolished when these neurons were infected with adenovirus encoding short hairpin RNA (shRNA) against FARP2 or mutant forms of FARP2.[47] Interestingly, the intracellular domain of plexins contains two highly conserved regions that share a high degree of homology to the GAP domain as well as containing two arginine residues that are essential for the catalytic activity, thus plexin itself may function as a GAP. Indeed, several lines of evidence demonstrated that Plexin-A1 and Plexin-B1 are GAPs for the Ras-family GTPase R-Ras and the activation of GAP activity of plexins leads to inactivation of R-Ras, resulting in detachment of cells from the extracellular matrix.[48,49] Together with the observation that stimulatory β1 integrin antibodies significantly block Sema3A-mediated growth cone repulsion,[47] these studies implicate the importance of membrane adhesion for Sema3A signaling and growth cone migration.

New Protein Synthesis

Local protein synthesis in the axon has been implicated in acute growth cone responses to several families of guidance cues, including Sema3A. Sema3A treatment resulted in a marked increase in protein synthesis in isolated growth cones of *Xenopus* retinal ganglion cells within minutes.[50] In addition, both Sema3A-induced growth cone collapse and repulsive turning were blocked by protein synthesis inhibitors. p42/p44 MAP kinase (MAPK) activated by Sema3A may be upstream to Sema3A-induced protein synthesis and subsequent chemotropic activity of growth cones.[51] Recently, it was shown that Sema3A induces intra-axonal translation of RhoA mRNA, and this local translation of RhoA is necessary and sufficient for Sema3A-mediated growth cone collapse.[52] β-actin is another potential candidate protein that is translated in response to guidance cues, since β-actin mRNA is transported down to growth cones[53] and disruption of β-actin mRNA and protein localization to the growth cone leads to impaired growth cone motility.[54]

Microdomain Signaling

Lipid rafts are plasma membrane microdomains enriched with cholesterol and glycosphingolipids, which provide an ordered lipid environment for localized trafficking and signaling.[55] A recent study showed that lipid rafts also mediate inhibitory effects of Sema3A on growth cones in both *Xenopus* spinal neurons.[56] Disruption of lipid rafts by membrane cholesterol depletion effectively blocks Sema3A-induced repulsion and extension of growth cones in *Xenopus* spinal neuron cultures.[56] In addition, a brief exposure to Sema3A increases the association of Neuropilin-1 with lipid rafts, implying asymmetric receptor-raft association and localized signaling in the growth cone during guidance responses. Activation of MAPK following

Sema3A treatment appears to depend on the integrity of lipid rafts and is required for Sema3A-induced growth cone repulsion.[51] These results support a role for lipid rafts in mediating growth cone guidance by providing a molecular platform for the localized assembly of ligand-receptor complex and their downstream effectors for cytoskeletal rearrangement and local protein synthesis, including Neuropilin-1, plexins, Src family kinases (SFKs), Rho-GTPases and MAPK.[55,56]

Modulation of Growth Cone Turning Responses to Class 3 Semaphorins

Cyclic nucleotides are potent modulators of growth cone turning in response to a group of guidance factors. For class 3 secreted semaphorin-induced growth cone responses, elevation of cyclic GMP (cGMP) signaling pathways converts Sema3A-induced repulsion of *Xenopus* spinal neurons into attraction in a growth cone turning assay.[18] Although cAMP analogs had no direct effect on Sema3A-induced repulsion, the cAMP antagonist Rp-cAMPs blocks the conversion of the turning response to Sema3A in the presence of 8-Br-cGMP, suggesting that there is some interaction between cAMP- and cGMP-dependent pathways.[18] The modulatory role of intracellular cyclic nucleotides in Sema3A-mediated repulsion was further supported by a recent report using a collagen gel assay showing that chemokine stromal cell-derived factor 1 (SDF-1) reduces the responsiveness of growth cones to Sema3A by elevating cAMP levels.[57] Interestingly, the level of cytoplamic cGMP seems to act as an endogenous regulator of Sema3A signaling because pharmacological and histological evidence suggests that asymmetric localization of soluble gunanylyl cyclase to the developing apical dendrites of cortical neurons allows Sema3A to act as a chemoattractant.[13]

Manipulation of the extracellular Ca^{2+} concentration, blockade of TRPC1-mediated Ca^{2+} influx or inhibition of the activity of CaM kinase II-calcineurin by specific inhibitors does not seem to influence Sema3A-induced growth cone repulsion of *Xenopus* spinal neurons.[18,58,59] Electrical activity stimulation and resultant Ca^{2+} influx, however, was shown to modulate Sema3A-induced growth cone guidance behaviors by enhancing the repulsive activity of Sema3A.[60] Since the enhanced repulsive effect by electrical stimulation is abolished either by the removal of extracellular Ca^{2+} or with the addition of Sp-cGMPs, a membrane-permeable analog of cGMP, and mimicked by Rp-cGMPs, a competitive analog of cGMP without electrical stimulation, it was proposed that electric stimulation may indirectly influence the growth cone responses by mechanisms involving cGMP pathways.[60] The molecular mechanisms of cross-talk among cAMP, cGMP and Ca^{2+} signaling are still elusive.

The functional cross-talk between cell adhesion protein L1 and Sema3A is implicated in repulsive responses to Sema3A.[61,62] L1, a member of the immunoglobulin superfamily of cell adhesion molecule (Ig CAM), directly associates with Neuropilin-1 and L1-deficient cortical and DRG axons lose their responsiveness to Sema3A, thereby acting as an integral part of the Sema3A-Neuropilin-1 receptor complex. L1 may also serve as a modulator of repulsive Sema3A signaling in two ways. First, L1 mediates the receptor internalization and thereby changes the sensitivity of growth cones to semaphorins.[63] Second, L1 may regulate the growth cone responses to Sema3A by decreasing the cGMP level. Addition of soluble L1-Fc chimera converted the Sema3A-mediated repulsion of wild-type but not L1-deficient axons into attraction through activation of NO/cGMP pathway.[61,62] On the other hand, blockade of soluble guanylate cyclase prevented the L1-Fc-induced switch in the Sema3A responses.

Summary

Despite a tremendous amount of progress in the identification and characterization of many new players as components of class 3 secreted semaphorin signaling in growth cone steering (Fig. 1), our understanding of the molecular mechanisms is far from complete. More questions remain to be answered: how are differential cytoskeletal changes within a growth cone achieved in response to semaphorins? What are the target(s) of cyclic nucleotide modulation? How does a growth cone make a reliable decision in response to a shallow gradient? And finally, how does a growth cone maintain its sensitivity to a decreasing concentration of semaphorins when it is growing away from the source? With a high degree of interest in the field with the development of novel technologies in analyzing growth cone steering, we expect to see a much more complete picture of semaphoring signaling in the near future.

References

1. Goodman CS TLM. Molecular mechanisms of axon guidance and target recognition. Molecular and cellular approaches to neural development 1997; 108-178.
2. Nakamura F, Kalb RG, Strittmatter SM. Molecular basis of semaphorin-mediated axon guidance. J Neurobiol 2000; 44(2):219-229.
3. Kruger RP, Aurandt J, Guan KL. Semaphorins command cells to move. Nat Rev Mol Cell Biol 2005; 6(10):789-800.
4. Luo Y, Raible D, Raper JA. Collapsin: A protein in brain that induces the collapse and paralysis of neuronal growth cones. Cell 1993; 75(2):217-227.
5. Fan J, Raper JA. Localized collapsing cues can steer growth cones without inducing their full collapse. Neuron 1995; 14(2):263-274.
6. Matthes DJ, Sink H, Kolodkin AL, Goodman CS. Semaphorin II can function as a selective inhibitor of specific synaptic arborizations. Cell 1995; 81(4):631-639.
7. Tanelian DL, Barry MA, Johnston SA et al. Semaphorin III can repulse and inhibit adult sensory afferents in vivo. Nat Med 1997; 3(12):1398-1401.
8. Taniguchi M, Yuasa S, Fujisawa H et al. Disruption of semaphorin III/D gene causes severe abnormality in peripheral nerve projection. Neuron 1997; 19(3):519-530.
9. Isbister CM, Tsai A, Wong ST et al. Discrete roles for secreted and transmembrane semaphorins in neuronal growth cone guidance in vivo. Development 1999; 126(9):2007-2019.
10. Bagnard D, Lohrum M, Uziel D et al. Semaphorins act as attractive and repulsive guidance signals during the development of cortical projections. Development 1998; 125(24):5043-5053.
11. Falk J, Bechara A, Fiore R et al. Dual functional activity of semaphorin 3B is required for positioning the anterior commissure. Neuron 2005; 48(1):63-75.
12. Polleux F, Giger RJ, Ginty DD et al. Patterning of cortical efferent projections by semaphorin-neuropilin interactions. Science 1998; 282(5395):1904-1906.
13. Polleux F, Morrow T, Ghosh A. Semaphorin 3A is a chemoattractant for cortical apical dendrites. Nature 2000; 404(6778):567-573.
14. Pasterkamp RJ, Kolodkin AL. Semaphorin junction: Making tracks toward neural connectivity. Curr Opin Neurobiol 2003; 13(1):79-89.
15. He Z, Wang KC, Koprivica V et al. Knowing how to navigate: Mechanisms of semaphorin signaling in the nervous system. Sci STKE 2002; 2002(119):RE1.
16. Huber AB, Kolodkin AL, Ginty DD et al. Signaling at the growth cone: Ligand-receptor complexes and the control of axon growth and guidance. Annu Rev Neurosci 2003; 26:509-563.
17. Lohof AM, Quillan M, Dan Y et al. Asymmetric modulation of cytosolic cAMP activity induces growth cone turning. J Neurosci 1992; 12(4):1253-1261.
18. Song H, Ming G, He Z et al. Conversion of neuronal growth cone responses from repulsion to attraction by cyclic nucleotides. Science 1998; 281(5382):1515-1518.
19. Messersmith EK, Leonardo ED, Shatz CJ et al. Semaphorin III can function as a selective chemorepellent to pattern sensory projections in the spinal cord. Neuron 1995; 14(5):949-959.
20. Puschel AW, Adams RH, Betz H. Murine semaphorin D/collapsin is a member of a diverse gene family and creates domains inhibitory for axonal extension. Neuron 1995; 14(5):941-948.
21. Polleux F, Ghosh A. The slice overlay assay: A versatile tool to study the influence of extracellular signals on neuronal development. Sci STKE 2002; 2002(136):PL9.
22. He Z, Tessier-Lavigne M. Neuropilin is a receptor for the axonal chemorepellent Semaphorin III. Cell 1997; 90(4):739-751.
23. Kolodkin AL, Levengood DV, Rowe EG et al. Neuropilin is a semaphorin III receptor. Cell 1997; 90(4):753-762.
24. Chen H, Chedotal A, He Z et al. Neuropilin-2, a novel member of the neuropilin family, is a high affinity receptor for the semaphorins Sema E and Sema IV but not Sema III. Neuron 1997; 19(3):547-559.
25. Takahashi T, Nakamura F, Jin Z et al. Semaphorins A and E act as antagonists of neuropilin-1 and agonists of neuropilin-2 receptors. Nat Neurosci 1998; 1(6):487-493.

26. Giger RJ, Urquhart ER, Gillespie SK et al. Neuropilin-2 is a receptor for semaphorin IV: Insight into the structural basis of receptor function and specificity. Neuron 1998; 21(5):1079-1092.
27. Winberg ML, Noordermeer JN, Tamagnone L et al. Plexin A is a neuronal semaphorin receptor that controls axon guidance. Cell 1998; 95(7):903-916.
28. Takahashi T, Fournier A, Nakamura F et al. Plexin-neuropilin-1 complexes form functional semaphorin-3A receptors. Cell 1999; 99(1):59-69.
29. Tamagnone L, Artigiani S, Chen H et al. Plexins are a large family of receptors for transmembrane, secreted, and GPI-anchored semaphorins in vertebrates. Cell 1999; 99(1):71-80.
30. Campbell DS, Regan AG, Lopez JS et al. Semaphorin 3A elicits stage-dependent collapse, turning, and branching in Xenopus retinal growth cones. J Neurosci 2001; 21(21):8538-8547.
31. Rohm B, Ottemeyer A, Lohrum M et al. Plexin/neuropilin complexes mediate repulsion by the axonal guidance signal semaphorin 3A. Mech Dev 2000; 93(1-2):95-104.
32. Cheng HJ, Bagri A, Yaron A et al. Plexin-A3 mediates semaphorin signaling and regulates the development of hippocampal axonal projections. Neuron 2001; 32(2):249-263.
33. Yaron A, Huang PH, Cheng HJ et al. Differential requirement for Plexin-A3 and -A4 in mediating responses of sensory and sympathetic neurons to distinct class 3 Semaphorins. Neuron 2005; 45(4):513-523.
34. Chedotal A, Del Rio JA, Ruiz M et al. Semaphorins III and IV repel hippocampal axons via two distinct receptors. Development 1998; 125(21):4313-4323.
35. Steup A, Lohrum M, Hamscho N et al. Sema3C and netrin-1 differentially affect axon growth in the hippocampal formation. Mol Cell Neurosci 2000; 15(2):141-155.
36. Pozas E, Pascual M, Nguyen Ba-Charvet KT et al. Age-dependent effects of secreted Semaphorins 3A, 3F, and 3E on developing hippocampal axons: In vitro effects and phenotype of Semaphorin 3A (-/-) mice. Mol Cell Neurosci 2001; 18(1):26-43.
37. Murakami Y, Suto F, Shimizu M et al. Differential expression of plexin-A subfamily members in the mouse nervous system. Dev Dyn 2001; 220(3):246-258.
38. Chen H, Bagri A, Zupicich JA et al. Neuropilin-2 regulates the development of selective cranial and sensory nerves and hippocampal mossy fiber projections. Neuron 2000; 25(1):43-56.
39. Mitsui N, Inatome R, Takahashi S et al. Involvement of Fes/Fps tyrosine kinase in semaphorin3A signaling. EMBO J 2002; 21(13):3274-3285.
40. Sasaki Y, Cheng C, Uchida Y et al. Fyn and Cdk5 mediate semaphorin-3A signaling, which is involved in regulation of dendrite orientation in cerebral cortex. Neuron 2002; 35(5):907-920.
41. Eickholt BJ, Walsh FS, Doherty P. An inactive pool of GSK-3 at the leading edge of growth cones is implicated in Semaphorin 3A signaling. J Cell Biol 2002; 157(2):211-217.
42. Terman JR, Mao T, Pasterkamp RJ et al. MICALs, a family of conserved flavoprotein oxidoreductases, function in plexin-mediated axonal repulsion. Cell 2002; 109(7):887-900.
43. Pasterkamp RJ, Dai HN, Terman JR et al. MICAL flavoprotein monooxygenases: Expression during neural development and following spinal cord injuries in the rat. Mol Cell Neurosci 2006; 31(1):52-69.
44. Mikule K, Gatlin JC, de la Houssaye BA et al. Growth cone collapse induced by semaphorin 3A requires 12/15-lipoxygenase. J Neurosci 2002; 22(12):4932-4941.
45. Luo L. Rho GTPases in neuronal morphogenesis. Nat Rev Neurosci 2000; 1(3):173-180.
46. Kubo T, Yamashita T, Yamaguchi A et al. A novel FERM domain including guanine nucleotide exchange factor is involved in Rac signaling and regulates neurite remodeling. J Neurosci 2002; 22(19):8504-8513.
47. Toyofuku T, Yoshida J, Sugimoto T et al. FARP2 triggers signals for Sema3A-mediated axonal repulsion. Nat Neurosci 2005; 8(12):1712-1719.
48. Oinuma I, Ishikawa Y, Katoh H et al. The Semaphorin 4D receptor Plexin-B1 is a GTPase activating protein for R-Ras. Science 2004; 305(5685):862-865.
49. Barberis D, Artigiani S, Casazza A et al. Plexin signaling hampers integrin-based adhesion, leading to Rho-kinase independent cell rounding, and inhibiting lamellipodia extension and cell motility. Faseb J 2004; 18(3):592-594.
50. Campbell DS, Holt CE. Chemotropic responses of retinal growth cones mediated by rapid local protein synthesis and degradation. Neuron 2001; 32(6):1013-1026.
51. Campbell DS, Holt CE. Apoptotic pathway and MAPKs differentially regulate chemotropic responses of retinal growth cones. Neuron 2003; 37(6):939-952.
52. Wu KY, Hengst U, Cox LJ et al. Local translation of RhoA regulates growth cone collapse. Nature 2005; 436(7053):1020-1024.
53. Zhang HL, Singer RH, Bassell GJ. Neurotrophin regulation of beta-actin mRNA and protein localization within growth cones. J Cell Biol 1999; 147(1):59-70.

54. Zhang HL, Eom T, Oleynikov Y et al. Neurotrophin-induced transport of a beta-actin mRNP complex increases beta-actin levels and stimulates growth cone motility. Neuron 2001; 31(2):261-275.
55. Tsui-Pierchala BA, Encinas M, Milbrandt J et al. Lipid rafts in neuronal signaling and function. Trends Neurosci 2002; 25(8):412-417.
56. Guirland C, Suzuki S, Kojima M et al. Lipid rafts mediate chemotropic guidance of nerve growth cones. Neuron 2004; 42(1):51-62.
57. Chalasani SH, Sabelko KA, Sunshine MJ et al. A chemokine, SDF-1, reduces the effectiveness of multiple axonal repellents and is required for normal axon pathfinding. J Neurosci 2003; 23(4):1360-1371.
58. Shim S, Goh EL, Ge S et al. XTRPC1-dependent chemotropic guidance of neuronal growth cones. Nat Neurosci 2005; 8(6):730-735.
59. Wen Z, Guirland C, Ming GL et al. A CaMKII/calcineurin switch controls the direction of Ca(2+)-dependent growth cone guidance. Neuron 2004; 43(6):835-846.
60. Ming G, Henley J, Tessier-Lavigne M et al. Electrical activity modulates growth cone guidance by diffusible factors. Neuron 2001; 29(2):441-452.
61. Castellani V, Chedotal A, Schachner M et al. Analysis of the L1-deficient mouse phenotype reveals cross-talk between Sema3A and L1 signaling pathways in axonal guidance. Neuron 2000; 27(2):237-249.
62. Castellani V, De Angelis E, Kenwrick S et al. Cis and trans interactions of L1 with neuropilin-1 control axonal responses to semaphorin 3A. EMBO J 2002; 21(23):6348-6357.
63. Piper M, Salih S, Weinl C et al. Endocytosis-dependent desensitization and protein synthesis-dependent resensitization in retinal growth cone adaptation. Nat Neurosci 2005; 8(2):179-186.

CHAPTER 6

Modulation of Semaphorin Signaling by Ig Superfamily Cell Adhesion Molecules

Ahmad Bechara, Julien Falk, Frédéric Moret and Valérie Castellani*

Summary

During axon navigation, growth cones continuously interact with molecular cues in their environment, some of which control adherence and bundle assembly, others axon elongation and direction. Growth cone responses to these different environmental cues are tightly coordinated during the development of neuronal projections. Several recent studies show that axon sensitivity to guidance cues is modulated by extracellular and intracellular signals. This regulation may enable different classes of cues to combine their effects and may also represent important means for diversifying pathway choices and for compensating for the limited number of guidance cues. This chapter focuses on the modulation exerted by Ig Superfamily cell adhesion molecules (IgSFCAMs) on guidance cues of the class III secreted semaphorins.

Introduction

In the developing nervous system, thousands of axon tracts form and are directed sometimes over long distances towards precise targets. The selective sorting and conveying of fiber tracts is controlled by interactions between receptor molecules expressed on axon terminals, the growth cones, and environmental cues. The different functional properties of these cues and regulation of their expression patterns are key mechanisms for the segregation of axon trajectories but other types of regulation might be mobilized to produce the tremendous diversity of pathways needed for elaborating neuronal connectivity. Analysis at the individual level revealed that growth cones equipped with functional receptors for guidance cues can display both repulsive and attractive responses, depending on intracellular and environmental contexts.[1] These observations strikingly overwhelmed the view that guidance cues can be categorized according to the nature of the effect they elicit. Importantly, they inferred the existence of regulations of guidance decisions at any point along axonal pathways. Intracellular modulation may enable axonal subpopulations expressing the same receptors to differently interpret a common set of environmental cues and therefore to select distinct pathways, while conversely, extra-cellular modulation may lead guidance cues expressed in different environments to elicit distinct effects. Both modes of regulation therefore represent important means for multiplying pathway choices. Importantly, such mechanisms may also enable extra-cellular cues with different functional properties to combine and coordinate their effects. Several recent studies

---

*Corresponding Author: Valérie Castellani—Centre de Génétique Moléculaire et Cellulaire (CGMC) UMR CNRS 5534, Université Claude Bernard, Lyon1, Campus de la Doua, Bâtiment Mendel 741, 43 Bd du 11 Novembre 1918, 69622 Villeurbanne cedex, France. Email: castellani@cgmc.univ-lyon1.fr

Semaphorins: Receptor and Intracellular Signaling Mechanisms, edited by R. Jeroen Pasterkamp. ©2007 Landes Bioscience and Springer Science+Business Media.

highlighted the influence of intracellular modulation in growth cone responses to guidance cues and are the topic of other chapters in this book. Here we focus on the regulation exerted by the Ig superfamily cell adhesion molecules (IgSFCAMs) on axonal responses to secreted semaphorin signals and discuss some biological contexts in which they may contribute to the guidance of axonal projections.

Soluble forms of IgSFCAMs Convert Repulsive Responses to Class III Semaphorins into Attraction

As detailed in other chapters, extensive searches for molecular cues controlling axon trajectory formation in the developing nervous system led to the cloning of a large axon guidance family, the semaphorins.[2,3] semaphorin members are subdivided into eight classes that share a 500 amino-acid "sema" domain, which is essential for their biological activity.[4] The Class III secreted semaphorins comprises seven members (from Sema3A to Sema3G,4-6), most of which are considered as major chemorepellents.[4] However these cues not only exert repulsive functions, because several axon sub-populations were attracted rather than repelled by sources of class III semaphorins.[7-10] Furthermore, growth cones can even change their response from repulsion to attraction, when the intracellular level of cyclic nucleotides is increased.[1]

Coincident extracellular signals also modulate semaphorin properties. In cocultures of cortical tissue with transfected COS7 cell aggregates, a soluble form of the IgSFCAM L1 could switch axonal responses to a class III semaphorin, Sema3A, from repulsion to attraction.[11] Such effect of soluble L1 was also observed when Sema3A was physiologically released from ventral spinal cord tissue. Functional links between IgSFCAMs and class III semaphorins were further evidenced by the recent finding that another IgSFCAM, NrCAM has similar properties to L1 but switches Sema3B and Sema3F signals from repulsion to attraction.[10] In other culture assays in which Sema3A, 3B or 3F was added to the medium of dissociated cell cultures, the growth cone collapsing effect normally exerted by these cues was prevented by coincident addition of soluble L1 or NrCAM. Thus, soluble forms of IgSFCAMs can modulate semaphorin signals.

L1, NrCAM, Neurofascin and CHL1 are transmembrane glycoproteins found in vertebrates that form, together with three invertebrate members, Neuroglian, tractin and LAD-1, a sub-group of IgSFCAMs expressed in developing axon tracts and migrating cells. L1 IgSFCAMs establish cis and trans homophilic and heterophilic interactions with IgSFCAM sub-groups, integrins and various extracellular matrix components such as chondroïtin sulfate proteoglycan.[12] This array of interactions enables L1 IgSFCAMs to mediate selective adhesive and anti-adhesive contacts and thus to play critical roles in cell migration, axon growth, fascicle formation and maintenance.[13-15] The extracellular part of L1 IgSFCAM proteins is composed of 6 Ig domains and 4 to 5 Fibronectin type III repeats. Their short and highly conserved cytoplasmic tail contains tyrosine and serine phosphorylation sites and binding motifs for members of the ankyrin and ERM cytoskeletal linked protein families (Fig. 1A).[16-17] Multiple variants of L1 IgSFCAMs are generated through alternative splicing and differential post-translational modifications such as glycosylation and proteolytic processing.[14] L1 IgSFCAMs can be released from the cell surface following cleavage by Matrix Metalloproteases (MMPs), A Disintegrin and Metalloprotease (Adam proteins), plasmin/plasminogen and furin convertase activities.[18-26] During the last five years interesting links have emerged between Adams, MMPs and axon guidance.[27] Adams are transmembrane proteins present in the growth cones and were shown to regulate axon behaviors through processing of guidance receptors and ligands.[28-29] These activities appear to be required at selected choice points along axon pathways. Pharmacological inhibition of MMP functions in *Xenopus* brain preparations, as well as genetic manipulations of *Drosophila* Adams result in axon guidance errors.[30-32] Soluble forms of IgSFCAMs are biologically active and IgSFCAM proteins extracted from various regions of the developing nervous system have different molecular weights, suggesting that transmembrane forms are processed in vivo.[12,33,34] Soluble L1 has also been detected in the developing brain where it might be released from environmental cell surfaces as well as from axons themselves.[21] Soluble IgSFCAMs are thus probably available for modulating semaphorin signals during axon navigation.

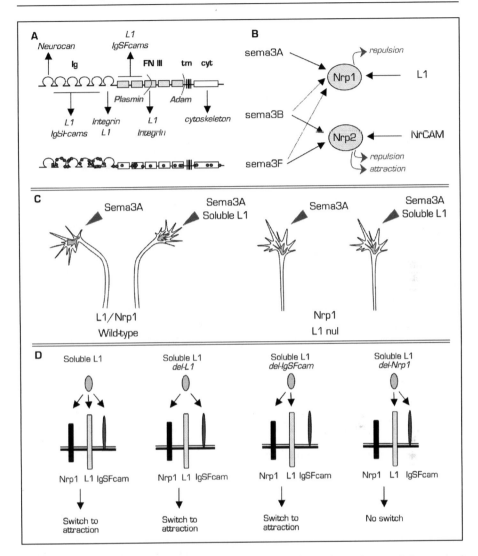

Figure 1. Functional interactions between IgSFCAMs and class III semaphorins. A) Structure of the IgSFCAM L1, binding partners and mutations. Functional studies of L1 mutations have established that residues of several Ig and FN domains of L1 are critical for homophilic and heterophilic binding.[68,69] B) Binding profile and requirement for IgSFCAMs in Semaphorin signaling. Sema3A only binds Nrp1 and requires Nrp1/L1 complex formation for mediating repulsion. Sema3B and Sema3F bind both Nrp1 and Nrp2, but they require Nrp2/NrCAM complex formation for mediating repulsion and attraction. C-D) Molecular interactions between IgSFCAMs and Nrps and growth cone responses to Semaphorins. C) Growth cones expressing Nrp1 and L1 are repelled by Sema3A and soluble L1 switches this response to attraction. Growth cones lacking L1 are not repelled by Sema3A and soluble L1 cannot elicit attraction: axons are insensitive to the cue. D) Soluble L1 can bind to L1, other IgSFCAMs and Nrp1 on the growth cone. Soluble L1 containing pathological mutations that interrupt L1 homophilic binding or binding to other IgSFCAMs, are still able to switch the Sema3A response. In contrast disruption of L1/Nrp1 binding prevents the effect of soluble L1, showing that Nrp1 is an obligatory receptor for soluble L1 in the modulation of Semaphorin signaling.

Transmembrane forms of IgCAMs Are Components of Class III Semaphorin Receptors

What are the molecular bases for the modulation of semaphorin signals by IgSFCAMs? Axonal receptors for class III semaphorins are multimolecular complexes formed by neuropilins (Nrp1 or Nrp2), which are the binding sub-units, and members of the Plexin-A (from A1 to A4) family, which transduce the signal.[35] Although Nrp/Plexin complexes are considered to be fully functional receptors, some contexts exist in which these complexes include additional molecular partners. As shown several years ago, the transmembrane form of L1 is also associated to Nrp1, the Nrp member utilized in the Sema3A receptor complex, and growth cones lacking L1 are insensitive to the repulsive effects exerted by Sema3A. The response to other class III semaphorins, which also bind Nrp1, was unaffected, suggesting a specific requirement for L1 in Sema3A signaling.[11] NrCAM associates with Nrp2 and is involved in axonal responses to Sema3F and Sema3B. In the latter case, it appears to be required for both attractive and repulsive effects, as shown by the use of NrCAM function-blocking antibodies in coculture experiments (Fig. 1A).[10]

IgSFCAM/Nrp complexes could be detected in precipitation experiments from transfected COS7 cells and from embryonic or neonatal brain extracts.[10-11] The interaction is constitutive and not disrupted in the presence of semaphorin ligands (Sema3A for L1/Nrp1 and Sema3F for NrCAM/Nrp2). Chimeric Fc-constructs of IgSFCAM ectodomains bind to Nrp expressing cells, indicating that these proteins associate by their extracellular domains. Finally, detection of semaphorin binding to transfected COS7 cells demonstrates that IgSFCAM/Nrp cis and trans interactions neither disrupt the semaphorin binding activity of Nrps nor their association with plexin A proteins.[10,11,36] Altogether, these finding are consistent with a model in which IgSFCAMs are molecular partners of Nrps in class III semaphorin receptor complexes.

Molecular Interactions Underlying the Modulation of Semaphorin Signaling by Soluble IgSFCAMs

How does the presence of an IgSFCAM in the semaphorin receptor relate with the properties of the corresponding soluble IgSFCAM to modulate the response to semaphorin signals? Insights into this question were given by considering a rare human disease caused by mutations in the gene encoding L1 on the X chromosome. L1 mutations are responsible for a spectrum of neurological disorders collectively referred to as X-linked hydrocephalus or MASA Syndrome.[16] About 150 mutations have been found, some of which lead to truncated proteins that do not reach the cell surface, disrupt L1 protein interactions or alter L1 expression level (Fig. 1A).[16]

To determine which L1 binding partner in the growth cone mediates the switch of Sema3A responses from repulsion to attraction, we sought to compare the functional properties of soluble mutated L1 either deficient for homophilic binding or for interactions with major IgSFCAM partners such as Tag-1/Axonin and F3/contactin.[12] Mutated L1 forms were added to cocultures of cortical explants and ventral spinal cord secreting Sema3A. Mutated forms that did not bind to L1 efficiently reversed axonal responses to Sema3A, as did those deficient for binding to Tag-1/axonin or F3/contactin. Further, L1 forms with increased binding affinities for these IgSFCAMs did not gain efficiency in switching the Sema3A signal.[36] This lack of correlation between the functional properties of L1 forms and their binding profile indicated that although interactions between soluble L1 and IgSFCAM partners present in the growth cones are likely to occur in a physiological context, such interactions are not instrumental for modulation of Sema3A signaling.

In contrast, one of the tested mutations, the L_{120V} affecting the first Ig1 domain of L1 did not alter L1/IgSFCAM interactions but totally blocked the reversion of Sema3A signal by soluble L1 (Fig. 1D). Further analysis showed that this mutation interrupts the binding of soluble L1 to Nrp1, suggesting that Nrp1 is the principal axonal receptor for soluble L1 involved in the reversion.[36] The use of these L1 mutated forms also allowed the mapping of the amino-acid sequence that mediates L1 binding to Nrp1. Several peptides were generated from the L1 sequence surrounding L_{120} and a five amino-acid peptide could mimic the switch by soluble L1, indicating that the L1/Nrp1 binding motif is present in the peptide. This motif also mediates L1/Nrp1 cis interaction, because transmembrane L1 with the L_{120V} mutation no longer coprecipitated with Nrp1.[36] Itoh and collaborators recently generated a knock-in mouse expressing a truncated form of L1, which lacks the Ig6 domain.[37] Surprisingly, although the deletion interrupts major L1 protein interactions, such as homophilic and heterophilic binding to IgSFCAMs and integrins, none of the guidance defects induced by L1 genetic ablation (which recapitulate those associated to the L1 human disease) were seen in the knock-in mouse model. Thus, the functions affected in the disease are preserved in this mouse and are not mediated by homophilic binding and association of L1 with integrins. Interestingly, neurons collected from the knock-in mice were found to still be responsive to Sema3A (ref. 37) a finding that provides additional support to the idea that some aspects of the L1 human disease could be due to alterations of functional links between L1 and Sema3A signaling. Another finding was that soluble L1 failed to elicit attractive responses of L1-deficient growth cones to Sema3A, indicating that the formation of L1/Nrp1 complex in cis is needed for the switch initiated by the binding of soluble L1 to Nrp1 in trans.[36] Thus, the modulation of Sema3A signaling occurs in specific contexts that enable appropriate conformation of soluble IgSFCAMs, their transmembrane counterparts and Nrps to assemble into macrocomplexes (Fig. 1).

Receptor Internalization and Modulation of Sema3A Signaling

As detailed in this paragraph, repulsive and attractive behaviors of axonal growth cones to Sema3A differ in the dynamics of their cell surface receptor expression. Sema3A (as well as other repellents) was shown to induce F-actin reorganization and endocytosis of receptor components during growth cone collapse.[38] This endocytosis appears to require the activity of the small RhoGTPase rac1 and to mediate fast desensitization of the growth cones to Sema3A exposure.[39-40] L1 was also found to be internalized and recycled at distinct domains of the growth cone during its progression on L1 substrates.[41-42] The rapid internalization of L1 is mediated by an alternatively spliced exon encoding the RSLE sequence, which creates a YRSL motif in the cytoplasmic domain of L1 that is recognized by the AP adaptor for clathrin dependant endocytosis.[41] This feature is specific to neuronal L1 forms as the exon is differentially spliced in other cell types such as Schwann cells.[43] Interestingly, cell contacts regulate L1 endocytosis via phosphorylation and dephosphorylation of the tyrosine residue in the motif, and this control of cell surface expression levels enables context-dependent modulation of L1 functions.[44]

The coupling of L1 to clathrin-dependant endocytic machinery is also utilized for internalizing Sema3A receptor components. Experiments performed on transfected COS7 cells indicated that Nrp1 and L1 are cointernalized by Sema3A treatment whereas Nrp1 remains at the cell surface, when expressed alone.[45] L1 forms lacking the YRSLE sequence or containing the L_{120V} mutation failed to internalize Nrp1. It is important to note that the capacity to internalize Nrp1 is probably not exclusive to L1 because Sema3A also induced endocytosis in Nrp1/plexin A1 expressing COS7 cells. Nevertheless in cultures of cortical neurons, Sema3A significantly increased endocytosis, as measured by Rhodamin-Dextran uptake, and also decreased L1 cell surface expression at the endocytic spots.[45] Such a link between L1 endocytosis and semaphorin signaling was also suggested by the finding that knock-down of CRMP-2, a member of the collapsin response mediated protein family, required for growth cone response to semaphorins (for details see other chapter), resulted in the inhibition of L1 endocytosis in cultured neurons.[46] Thus, Sema3A induces internalization of plexin A, L1 and Nrp1, but it remains to be determined whether L1 or plexin A mediates this process.

Strikingly, in similar types of experiments, the modulation of Sema3A signal by soluble L1 was correlated with a blockade of L1/Nrp1 internalization.[45] In transfected COS7 cells treated with a combination of Sema3A and soluble L1 or its mimetic peptide, Nrp1/L1 surface expression was maintained. The same treatment applied to cortical neurons collectively blocked the growth cone collapse, the Rhodamin-dextran uptake and the reduction of L1 cell surface localization.[45] Thus, differences in the localization or the dynamics of the Sema3A receptor may underlie the opposing growth cone responses, with receptor internalization associated with repulsion but not with attraction.

What could be the functional consequences of different receptor dynamics for the growth cone? First, internalized and cell surface receptors could be coupled to different signaling machinery. Alternatively, similar cascades could be initially activated but at distinct levels and/or duration, leading to bifurcation downstream in the signaling cascade. Interesting links are effectively made between signaling cascade components and ligand/receptor sub-cellular compartmentalization, and endocytosis not only serves for termination of the signal.[47-48] For example, the MAPK pathway is often (but not exclusively) activated upon receptor endocytosis. MAPK activation triggered by L1, which mediates the growth-promoting properties of the cue, is indeed subordinated to L1 internalization.[49] Phosphorylation of Erk1/2 MAPK effectors is also required for growth cone responses to Sema3A, as shown by the use of pharmacological inhibitors in *Xenopus* retinal cultures.[50] It remains unclear whether MAPK activation depends on receptor internalization in this context, but if so, modifications of receptor internalization would certainly modulate it. Differences in the signaling cascades controlling repulsive and attractive behaviors have been illustrated by the finding that stimulation of nitric oxide (NO) and cGMP production switches Sema3A repulsion to attraction. Whether differences in receptor dynamics are able to control the levels of these messengers is however not known.[1,36] Second, regulated endocytosis could be part of the mechanisms by which the adherence of growth cone structures is adjusted to modifications of growth cone direction. Internalization of L1 could result in a local decrease of adhesion, which could be eventually amplified by homomultimerization and linkage of L1 to other adhesion systems such as integrins.[20] Removal of adherence forces could facilitate withdrawal of the growth cone membrane. Conversely, by blocking L1 internalization, which was shown to result in increased adherence within minutes (ref. 51), soluble L1 could stabilize the growth cone structures to favor attractive behavior (Fig. 2).

Biological Contexts for Regulation of Semaphorin Signals by IgSFCAMS

A requirement for IgSFCAMs in growth cone responses to semaphorins has been suggested for several classes of axons. Initially, ventral Sema3A expression was observed to delineate the boundary between the caudal medulla and the spinal cord, a region in which axons descending ventrally from the cerebral cortex, cross the midline towards the dorsal funiculus, and then project to contralateral spinal motoneurons and interneurons. Corticospinal axons from wild-type but not L1-deficient mice were repelled by this ventral source of Sema3A.[11] This observation was consistent with a previous analysis of L1 knockout mice, which showed abnormal ventral growth of axons at the midline crossing level.[52] NrCAM was found to be required for the response of axons forming another major commissural projection, the anterior commissure, which interconnects olfactory and temporal cortical structures of the brain.[10] Defects of the anterior commissure were found in Sema3B and NrCAM null mutants as well as in Sema3F and Nrp2 knockouts.[10,53-55] Thus, modulation of the sensitivity to semaphorins may be perhaps needed for commissural axons to take appropriate guidance decisions before, during or after midline crossing.

Regulation of Pathway Choices at the Midline?

Midline crossing is a highly complex process, which has been studied extensively using the model of spinal commissural projections.[56] It requires a dynamic regulation of growth cone sensitivity to long-range and short-range midline-derived chemo-attractants and repellents.[56] Several protein families were found to control guidance decisions at the midline including class III semaphorins, as revealed by the defects of midline crossing observed in Nrp2 null mutant embryos.[57] Notably, L1 IgSFCAMs also critically regulate

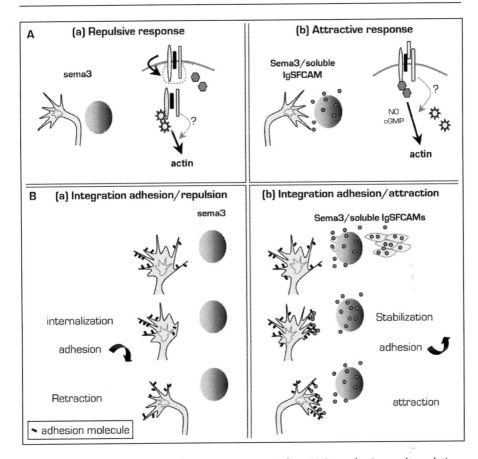

Figure 2. Direction of the growth cone response to class III Semaphorins and regulations triggered by IgSFCAMs. A) Sub-cellular localization of the Sema receptor and activation of signaling pathways. In (a) Endocytosis of the receptor leads to activation of signaling pathways mediating a repulsive behavior. In (b) the receptor is not internalized and triggers activation of a different cascade, which includes NO and cGMP production, and mediates an attractive behavior. The question mark illustrates the possible contribution of the IgSFCAM to signal transduction mechanisms. B) Functional coupling of mechanisms controlling adhesion and guidance: a possible role for IgSFCAMs. a) Coendocytosis of Nrp and IgSFCam in growth cone structures facing the Sema3 source is higher and this asymmetric internalization of IgSFCAMs generates a polarized decrease of adhesiveness, which facilitates morphological growth cone retraction and turning. b) Soluble IgSFCams inhibit receptor endocytosis, and by maintaining cell surface expression of adhesion molecules, create a polarized increase of adhesiveness which stabilizes the growth cone structures facing the source of Semaphorin and favors attractive behavior. In both cases, adhesive features of the growth cone are controlled by the level of IgSFCAM cell surface expression, and are adapted to the direction of the growth cone response.

midline crossing by spinal commissural axons. Injection of function blocking antibodies or electroporation of siRNAs to L1 IgSFCAM transcripts in the chick embryo resulted in striking defects of midline crossing.[58,59] Sema3B and Sema3F are synthesized by midline or surrounding cells of the ventral spinal cord and interestingly, commissural axons gain a repulsive response to these cues at the post-crossing stage.[57] Modulation of semaphorin responses by IgSFCAMs would certainly make sense in such situations where growth cones alter, modify or gain sensitivity to environmental cues in a short amount of time and limited space. Although this remains speculative, fast and reversible modifications of IgSFCAM cell surface expression, achieved by regulated-endocytosis or locally activated proteolytic processing leading to IgSFCAM removal and extra-cellular release, could participate in the modulation of growth cone sensitivity to semaphorin signals occurring at the midline (Fig. 3).

Waiting Periods in Target Inervation?

Another possible context for an IgSFCAM/semaphorin interplay is the temporal pattern of target innervation. Waiting periods during which afferents are prevented to enter their targets have been described for several systems of projections and assumed to allow maturational changes to occur in the target field.[60-62] In the chick embryo, sensory and motor axons wait in the plexus at the base of the limb for about one day, and during this period, axon fascicles destined for common target muscles are organized.[60,62] L1 and PSA-NCAM mediate important signals for the selective rearrangement of motor axon tracts and subsequent pathway choices in the limb. These cues were also suggested to modulate the interactions between axons and guidance cues in the plexus.[63] Interestingly, recent work showed that motor axons prematurely invade the limb in Sema3A null mutant mice, suggesting that Sema3A is one of the cues controlling waiting periods and in-growth to target fields.[64]

Innervation of target cells is often achieved by emission of collaterals from primary axons, as is the case for central afferents of DRG sensory neurons. These initially grow towards the dorsal root entry zone (DREZ) and extend longitudinally, to avoid entering the neural tube. Collaterals are then formed, which select specific pathways to innervate appropriate grey matter layers, depending on their sensory modality. Roles for both Sema3A and IgSFCAMs were proposed in the timing and patterning of spinal innervation. In ovo injection of function blocking antibodies to IgSFCAMs induced premature entry in the dorsal horn and pathfinding errors. L1/axonin-1 interactions were found required for nociceptive innervation of the dorsal horn, and contactin/NrCAM interactions for proprioceptive innervation of the ventral horn.[65] Interestingly, repulsion by Sema3A was also proposed to restrict the early growth of sensory afferents in the DREZ, and later to prevent nociceptive afferents from innervating the ventral horn.[66-67] Thus, modifications of growth cone sensitivity to semaphorins could be part of the processes leading to premature entry and pathfinding errors due to IgSFCAM function blocking antibodies. Altogether, these data bring several contexts in which IgSFCAM and semaphorin/Neuropilin signaling could function in common mechanisms for regulating waiting periods and pathway choices.

Conclusions

The chemorepulsive properties of class III semaphorins have been extensively documented and shown to contribute to the guidance of both central and peripheral nerve projections. Although fewer studies illustrate attractive behaviors to semaphorins, undoubtedly this type of response also contributes to axonal pathfinding. Whether switch of axon responses to semaphorins is indeed indispensable during axon navigation is much more hypothetical, but unquestionably, the bi-functional potential of guidance cues provides a means for amplifying the number of pathway choices dictated by limited sets of signaling molecules. The recruitment of IgSFCAMs to semaphorin receptors may be part of the mechanisms by which adhesiveness of growth cone structures is adjusted during changes in axon direction. It may also allow the bipotential nature of growth cone responses to semaphorins to be exerted in line with coincident environmental cues encountered along axonal pathways. Many questions remain unanswered. Where and when is such modulation of growth cone sensitivity exerted? What are the intracellular molecular pathways? Are other IgSFCAMs able to regulate semaphorin signals? Insights should come with the characterization of signaling events specifically implicated in attractive and repulsive behaviors.

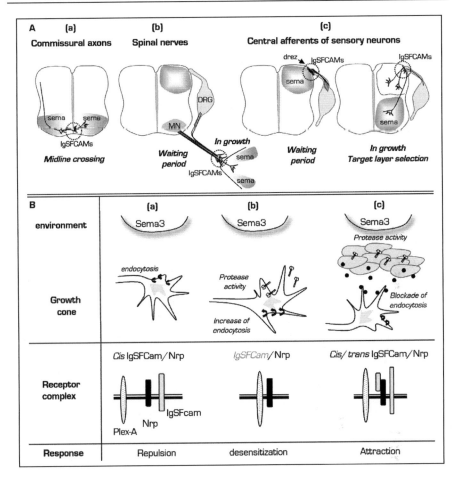

Figure 3. IgSFCAMs and axonal responses to class III Semaphorins. A) Guidance decisions and pathway choices that could involve modulation of growth cone responses to Semaphorins by IgSFCAMs. a) Spinal commissural axons grow towards the ventral floor plate in which they cross the midline. The regulation of growth cone sensitivity to Semaphorins and interactions involving IgSFCAMs both contribute to guidance decisions at the midline. b) Motor (MN) and DRG sensory axons are arrested at the base of the limb. During this waiting period, axons defasciculate and rearrange according to the identity of their target muscles. Both IgSFCAMs and Semaphorins contribute to the timing of the waiting period and the patterning of muscle innervation. (c) Central afferences extended by DRG sensory neurons wait in the dorsal root entry zone (DREZ), prior to emit collaterals which select distinct pathways for innervation of grey matter layers. Both IgSFCAMs and Semaphorin (Sema3A) were suggested to control the waiting period in the DREZ and the segregation of projections destined for ventral and dorsal innervation. B) Processing of IgSFCAMs that could influence the nature of growth cone responses to Semaphorins. a) IgSFCAMs are complexed to Nrps: the growth cone has a repulsive behavior, during which the complex is internalized. b) Modification of growth cone features, characterized by increased endocytosis or activation of intrinsic or extrinsic protease activity, leads to removal of IgSFCam cell surface expression. The IgSFCAM/Nrp complex is disrupted: the growth cone is desensitized. C) Modification of environmental features, such as an activation of protease activity, leads to the release of soluble IgSFCAMs, which bind to Nrps and block the receptor endocytosis: the growth cone switches its response from repulsion into attraction. Such regulation of growth cone sensitivity could contribute to guidance decisions in the contexts described in (A).

Acknowledgments

We thank Edmund Derrington for helpful comments on the manuscript. This work was funded by grants from the CNRS ATIP program (Centre National de la recherché Scientifique) FRM (Fondation pour la Recherche Médicale) and AFM (Association Française contre les Myopathies).

References

1. Song HJ, Poo MM. Signal transduction underlying growth cone guidance by diffusible factors. Curr Opin Neurobiol 1999; (3):355-63.
2. Luo Y, Raible D, Raper JA. Collapsin: A protein in brain that induces the collapse and paralysis of neuronal growth cones. Cell 1993; 75(2):217-27.
3. Kolodkin AL, Matthes DJ, Goodman CS. The semaphorin genes encode a family of transmembrane and secreted growth cone guidance molecules. Cell 1993; 75(7):1389-9.
4. Raper JA. semaphorins and their receptors in vertebrates and invertebrates. Curr Opin Neurobiol 2000; 10(1):88-94.
5. Stevens CB, Halloran MC. Developmental expression of sema3G, a novel zebrafish semaphorin. Gene Expr Patterns 2005; 5(5):647-53.
6. Taniguchi M, Masuda T, Fukaya M et al. Identification and characterization of a novel member of murine semaphorin family. Genes Cells 2005; 10(8):785-92.
7. Bagnard D, Lohrum M, Uziel D et al. semaphorins act as attractive and repulsive guidance signals during the development of cortical projections. Development 1998; 125:5043-5053.
8. de Castro F, Hu L, Drabkin H et al. Chemoattraction and chemorepulsion of olfactory bulb axons by different secreted semaphorins. J Neurosci 1999; 19:4428-36.
9. Wolman MA, Liu Y, Tawarayama H et al. Repulsion and attraction of axons by semaphorin3D are mediated by different neuropilins in vivo. J Neurosci 2004; 24:8428-8435.
10. Falk J, Bechara A, Fiore R et al. Dual functional activity of semaphorin 3B is required for positioning the anterior commissure. Neuron 2005; 48(1):63-75.
11. Castellani V, Chedotal A, Schachner M et al. Analysis of the L1-deficient mouse phenotype reveals cross-talk between Sema3A and L1 signaling pathways in axonal guidance. Neuron 2000; 27:237-249.
12. Brummendorf T, Rathjen FG. Structure/function relationships of axon-associated adhesion receptors of the immunoglobulin superfamily. Curr Opin Neurobiol 1996; 6(5):584-93.
13. Kamiguchi H, Lemmon V. IgCAMs: Bidirectional signals underlying neurite growth. Curr Opin Cell Biol 2000; 12(5):598-60.
14. Brummendorf T, Lemmon V. Immunoglobulin superfamily receptors: Cis-interactions, intracellular adapters and alternative splicing regulate adhesion. Curr Opin Cell Biol 2001; 13:611.
15. Rougon G, Hobert O. New insights into the diversity and function of neuronal immunoglobulin superfamily molecules. Annu Rev Neurosci 2003; 26:207-38.
16. Brummendorf T, Kenwrick S, Rathjen FG. Neural cell recognition molecule L1: From cell biology to human hereditary brain malformations. Curr Opin Neurobiol 1998; 8(1):87-97.
17. Dickson TC, Mintz CD, Benson DL et al. Functional binding interaction identified between the axonal CAM L1 and members of the ERM family. J Cell Biol 2002; 24:1105-12.
18. Volkmer H, Hassel B, Wolff JM et al. Structure of the axonal surface recognition molecule neurofascin and its relationship to a neural subgroup of the immunoglobulin superfamily. J Cell Biol 1992; 118(1):149-61.
19. Gutwein P, Oleszewski M, Mechtersheimer S et al. Role of Src kinases in the ADAM-mediated release of L1 adhesion molecule from human tumor cells. J Biol Chem 2000; 275(20):15490-7.
20. Silletti S, Mei F, Sheppard D et al. Plasmin-sensitive dibasic sequences in the third fibronectin-like domain of L1-cell adhesion molecule (CAM) facilitate homomultimerization and concomitant integrin recruitment. J Cell Biol 2000; 149(7):1485-50.
21. Mechtersheimer S, Gutwein P, Agmon-Levin N et al. Ectodomain shedding of L1 adhesion molecule promotes cell migration by autocrine binding to integrins. J Cell Biol 2001; 155(4):661-73.
22. Gutwein P, Mechtersheimer S, Riedle S et al. ADAM10-mediated cleavage of L1 adhesion molecule at the cell surface and in released membrane vesicles. FASEB J 2003; 17(2):292-4.
23. Xu YZ, Ji Y, Zipser B et al. Proteolytic cleavage of the ectodomain of the L1 CAM family member Tractin. J Biol Chem 2003; 278(6):4322-30.
24. Naus S, Richter M, Wildeboer D et al. Ectodomain shedding of the neural recognition molecule CHL1 by the metalloprotease-disintegrin ADAM8 promotes neurite outgrowth and suppresses neuronal cell death. J Biol Chem 2004; 279(16):16083-9.

25. Conacci-Sorrell M, Kaplan A, Raveh S et al. The Shed Ectodomain of Nr-CAM stimulates cell proliferation and motility, and confers cell transformation. Cancer Res 2005; 65(24):11605-12.
26. Maretzky T, Schulte M, Ludwig A et al. L1 is sequentially processed by two differently activated metalloproteases and Presenilin/gamma-secretase and regulates neural cell adhesion, cell migration, and neurite outgrowth. Mol Cell Biol 2005; 25(20):9040-53.
27. McFarlane S. Metalloproteases: Carving out a role in axon guidance. Neuron 2003; 37(4):559-62.
28. Hattori M, Osterfield M, Flanagan JG. Regulated cleavage of a contact-mediated axon repellent. Science 2000; 289(5483):1360-5.
29. Galko MJ, Tessier-Lavigne M. Function of an axonal chemoattractant modulated by metalloprotease activity. Science 2000; 289(5483):1365-7.
30. Fambrough D, Pan D, Rubin GM et al. The cell surface metalloprotease/disintegrin Kuzbanian is required for axonal extension in Drosophila. Proc Natl Acad Sci USA 1996; 93(23):13233-8.
31. Schimmelpfeng K, Gogel S, Klambt C. The function of leak and kuzbanian during growth cone and cell migration. Mech Dev 2001; 106(1-2):25-36.
32. Hehr CL, Hocking JC, McFarlane S. Matrix metalloproteinases are required for retinal ganglion cell axon guidance at select decision points. Development 2005; 132(15):3371-9.
33. Kayyem JF, Roman JM, de la Rosa EJ et al. Bravo/Nr-CAM is closely related to the cell adhesion molecules L1 and Ng-CAM and has a similar heterodimer structure. J Cell Biol 1992; 118(5):1259-7.
34. Volkmer H, Hassel B, Wolff JM et al. Structure of the axonal surface recognition molecule neurofascin and its relationship to a neural subgroup of the immunoglobulin superfamily. J Cell Biol 1992; 118(1):149-61.
35. Tamagnone L, Comoglio PM. Signaling by semaphorin receptors: Cell guidance and beyond. Trends Cell Biol 2000; 10:377-383.
36. Castellani V, De Angelis E, Kenwrick S et al. Cis and trans interactions of L1 with neuropilin-1 control axonal responses to semaphorin 3A. EMBO J 2002; 21:6348-6357.
37. Itoh K, Cheng L, Kamei Y et al. Brain development in mice lacking L1-L1 homophilic adhesion. J Cell Biol 2004; 165(1):145-54.
38. Fournier AE, Nakamura F, Kawamoto S et al. semaphorin3A enhances endocytosis at sites of receptor-F-actin colocalization during growth cone collapse. J Cell Biol 2000; 149(2):411-22.
39. Jurney WM, Gallo G, Letourneau PC et al. Rac1-mediated endocytosis during ephrin-A2- and semaphorin 3A-induced growth cone collapse. J Neurosci 2002; 22(14):6019-28.
40. Piper M, Salih S, Weinl C et al. Endocytosis-dependent desensitization and protein synthesis-dependent resensitization in retinal growth cone adaptation. Nat Neurosci 2005; 8(2):179-86.
41. Kamiguchi H, Long KE, Pendergast M et al. The neural cell adhesion molecule L1 interacts with the AP-2 adaptor and is endocytosed via the clathrin-mediated pathway. J Neurosci 1998; 18(14):5311-21.
42. Kamiguchi H, Lemmon V. Recycling of the cell adhesion molecule L1 in axonal growth cones. J Neurosci 2000; 20(10):3676-86.
43. Reid RA, Hemperly JJ. Variants of human L1 cell adhesion molecule arise through alternate splicing of RNA. J Mol Neurosci 1992; 3(3):127-35.
44. Schaefer AW, Kamei Y, Kamiguchi H et al. L1 endocytosis is controlled by a phosphorylation-dephosphorylation cycle stimulated by outside-in signaling by L1. J Cell Biol 2002; 157(7):1223-32.
45. Castellani V, Falk J, Rougon G. semaphorin3A-induced receptor endocytosis during axon guidance responses is mediated by L1 CAM. Mol Cell Neurosci 2004; 26(1):89-10.
46. Nishimura T, Fukata Y, Kato K et al. CRMP-2 regulates polarized Numb-mediated endocytosis for axon growth. Nat Cell Biol 2003; 5:819-26.
47. McPherson PS, Kay BK, Hussain NK. Signaling on the endocytic pathway. Traffic 2001; 2(6):375-84.
48. Kholodenko BN. Four-dimensional organization of protein kinase signaling cascades: The roles of diffusion, endocytosis and molecular motors. J Exp Biol 2003; 206(Pt 12):2073-82.
49. Schaefer AW, Kamiguchi H, Wong EV et al. Activation of the MAPK signal cascade by the neural cell adhesion molecule L1 requires L1 internalization. J Biol Chem 1999; 274(53):37965-7.
50. Campbell DS, Holt CE. Apoptotic pathway and MAPKs differentially regulate chemotropic responses of retinal growth cones. Neuron 2003; 37(6):939-52.
51. Long KE, Asou H, Snider MD et al. The role of endocytosis in regulating L1-mediated adhesion. J Biol Chem 2001; 276(2):1285-9.
52. Cohen NR, Taylor JS, Scott LB et al. Errors in corticospinal axon guidance in mice lacking the neural cell adhesion molecule L1. Curr Biol 1998; 8(1):26-3.
53. Chen H, Bagri A, Zupicich JA et al. Neuropilin-2 regulates the development of selective cranial and sensory nerves and hippocampal mossy fiber projections. Neuron 2000; 25:43-56.

54. Giger RJ, Cloutier JF, Sahay A et al. Neuropilin-2 is required in vivo for selective axon guidance responses to secreted semaphorins. Neuron 2000; 25:29.
55. Sahay A, Molliver ME, Ginty DD et al. semaphorin 3F is critical for development of limbic system circuitry and is required in neurons for selective CNS axon guidance events. J Neurosci 2003; 23:6671-6680.
56. Kaprielian Z, Runko E, Imondi R. Axon guidance at the midline choice point. Dev Dyn 2001; 221(2):154-81.
57. Zou Y, Stoeckli E, Chen H et al. Squeezing axons out of the gray matter: A role for slit and semaphorin proteins from midline and ventral spinal cord. Cell 2000; 102:363-375.
58. Stoeckli ET, Landmesser LT. Axonin-1, Nr-CAM, and Ng-CAM play different roles in the in vivo guidance of chick commissural neurons. Neuron 1995; 14(6):1165-79.
59. Pekarik V, Bourikas D, Miglino N et al. Screening for gene function in chicken embryo using RNAi and electroporation. Nat Biotechnol 2003; 21(1):93-6.
60. Tosney KW, Landmesser LT. Development of the major pathways for neurite outgrowth in the chick hindlimb. Dev Biol 1985; 109(1):193-214.
61. Ghosh A, Shatz CJ. Pathfinding and target selection by developing geniculocortical axons. J Neurosci 1992; 12(1):39-55.
62. Wang G, Scott SA. The "waiting period" of sensory and motor axons in early chick hindlimb: Its role in axon pathfinding and neuronal maturation. J Neurosci 2000; 20(14):5358-66.
63. Tang J, Rutishauser U, Landmesser L. Polysialic acid regulates growth cone behavior during sorting of motor axons in the plexus region. Neuron 1994; 13(2):405-14.
64. Huber AB, Kania A, Tran TS et al. Distinct roles for secreted semaphorin signaling in spinal motor axon guidance. Neuron 2005; 48(6):949-64.
65. Perrin FE, Rathjen FG, Stoeckli ET. Distinct subpopulations of sensory afferents require F11 or axonin-1 for growth to their target layers within the spinal cord of the chick. Neuron 2001; 30(3):707-23.
66. Wright DE, White FA, Gerfen RW et al. The guidance molecule semaphorin III is expressed in regions of spinal cord and periphery avoided by growing sensory axons. J Comp Neurol 1995; 361(2):321-33.
67. Fu SY, Sharma K, Luo Y et al. SEMA3A regulates developing sensory projections in the chicken spinal cord. J Neurobiol 2000; 45(4):227-36.
68. De Angelis E, Watkins A, Schafer M et al. Disease-associated mutations in L1 CAM interfere with ligand interactions and cell-surface expression. Hum Mol Genet 2002; 11(1):1-1.
69. De Angelis E, MacFarlane J, Du JS et al. Pathological missense mutations of neural cell adhesion molecule L1 affect homophilic and heterophilic binding activities. EMBO J 1999; 18(17):4744-53.

CHAPTER 7

Proteoglycans as Modulators of Axon Guidance Cue Function

Joris de Wit* and Joost Verhaagen

Abstract

Organizing a functional neuronal network requires the precise wiring of neuronal connections. In order to find their correct targets, growth cones navigate through the extracellular matrix guided by secreted and membrane-bound molecules of the slit, netrin, ephrin and semaphorin families. Although many of these axon guidance molecules are able to bind to heparan sulfate proteoglycans, the role of proteoglycans in regulating axon guidance cue function is only now beginning to be understood. Recent developmental studies in a wide range of model organisms have revealed a crucial role for heparan sulfate proteoglycans as modulators of key signaling pathways in axon guidance. In addition, emerging evidence indicates an essential role for chondroitin sulfate proteoglycans in modifying the guidance function of semaphorins. It is becoming increasingly clear that extracellular matrix molecules, rather than just constituting a structural scaffold, can critically influence axon guidance cue function in development, and may continue to do so in the injured adult nervous system.

Introduction

Correct wiring of neuronal connections requires precise steering of growth cones. Axonal growth cones are instructed by secreted and membrane-bound guidance molecules as they navigate through the extracellular matrix (ECM) towards distant targets. These axon guidance cues can act as growth cone attractants or repellents, and can be short-range (membrane-attached) or long-range (diffusible) in nature.[1] Four major families of axon guidance cues have been identified; the slits, netrins, ephrins and semaphorins.[2] An important constituent of the ECM are the proteoglycans, cell surface and ECM molecules consisting of a core protein and co-valently attached glycosaminoglycan (GAG) sidechains. These highly sulfated GAG sidechains are made up of large polymers of repeating disaccharide units and can be grouped in two major classes, heparan sulfate (HS) and chondroitin sulfate (CS) chains, depending on the disaccharide structure. A large variety of HSPGs and CSPGs is found in the developing and mature nervous sytem.[3]

Once considered to merely act as structural components of the ECM, proteoglycans are now widely recognized as essential modulators of ligand-receptor interactions in key signaling pathways in many developmental processes.[4] Secreted proteins of the fibroblast growth factor (FGF), Wnt, transforming growth factor β (TGF-β) and Hedgehog families have been shown to be functionally dependent on the presence of HSPGs. HSPGs can regulate the activities of

*Corresponding Author: Joris de Wit—Department of Functional Genomics, Center for Neurogenomics and Cognitive Research, De Boelelaan 1087, 1081 HV Amsterdam, The Netherlands. Email: joris@cncr.vu.nl; or Joost Verhaagen, j.verhaagen@nih.knaw.nl

Semaphorins: Receptor and Intracellular Signaling Mechanisms, edited by R. Jeroen Pasterkamp.
©2007 Landes Bioscience and Springer Science+Business Media.

these developmental signaling pathways by controlling the extracellular distribution of secreted ligands and by facilitating ligand-receptor interactions.[5,6] Although most research has thus far focused on the role of HSPGs in development, recent insights suggest an equally important role for CSPGs.[7]

While many axon guidance proteins were biochemically purified on the basis of their heparin binding properties,[8-10] the functional role of HSPGs in regulating axon pathfinding is only now beginning to be understood. Recent in vitro and genetic studies in model systems ranging from worm to mouse have illustrated the importance of HSPGs in modulating axon guidance cue function and point to a similar role for CSPGs as well. Here, we will review the evidence for a role of proteoglycans in regulating the activity and distribution of secreted and transmembrane proteins of the major axon guidance cue families in nervous system development and speculate on the role these interactions might play in the regeneration and plasticity of adult injured neurons.

Role of Heparan Sulfate Proteoglycans in Axon Guidance

The binding of secreted guidance cues to proteoglycans in the ECM can serve a role in the localization of soluble proteins, thereby establishing gradients or defining boundaries to provide spatial information for navigating growth cones, but might also be of relevance for their biological activity. Initial indications for the involvement of HSPGs in axon guidance came from studies that analyzed axon tract development following experimental manipulation of HSPGs. The addition of exogenous HS or the enzymatic removal of HS resulted in defasciculation and guidance defects of specific pioneer axon tracts in cultured cockroach embryos.[11] Similarly, exogenous HS treatment or enzymatic digestion of HSPGs resulted in profound retinal axon growth and guidance defects in the *Xenopus* visual system.[12,13] The observed axon guidance defects following experimental manipulation of HSPGs suggest that guidance cues functionally associated with HSPGs are competed away by exogenous HS treatment or displaced following enzymatic removal of HS. Recent work on the Slit family of guidance cues has provided compelling evidence for a functional association of secreted guidance molecules with HSPGs. In addition, emerging evidence indicates a role for proteoglycans in axon guidance mediated by the transmembrane semaphorin Sema5A. Below, we will discuss the role of HSPGs in axon guidance mediated by secreted and transmembrane proteins of the Slit, netrin, semaphorin and ephrin gene families.

Slits and Heparan Sulfate Proteoglycans

Biochemical studies have shown that Slit proteins bind to HSPGs. The mammalian Slit2 protein was purified from postnatal brain membrane fractions using heparin columns.[9] Upon expression in heterologous cells, secreted Slit2 is tightly bound to cell surfaces from which it can be removed with a heparin wash.[14] Slit2 binds to the GPI-anchored HSPG glypican-1, and this binding is decreased by competition with excess heparin or enzymatic removal of HS chains.[15,16] Examination of the Slit protein distribution in *Drosophila* has shown that Slit is mainly associated with the surface of Slit-producing midline glial cells and can also be detected on the surface of commissural axons crossing these cells.[17,18] Slit2 protein associates with the surface of reactive astrocytes in the injured rodent cerebral cortex,[19] suggesting a limited diffusion of Slit proteins in tissue.

Recent evidence has shown that the binding of Slit proteins to HS is essential for their biological activity. Enzymatic removal of cell surface HS by heparinase treatment decreased the affinity of Slit2 for its receptor Robo and abolished the repulsive activity of Slit2 on olfactory axons, showing that HS is required for Slit2-mediated repulsion in vitro.[20] Similarly, a more recent study showed that the repulsive effect of Slit2 on *Xenopus* retinal axons is lost when these axons are treated with exogenous HS or heparinase.[21] These in vitro results are supported by recent genetic evidence from zebrafish, worm, fly and mouse, which demonstrated a functional role for HS in modulating Slit-mediated axon guidance at the midline.

Conditional disruption of the HS-polymerizing enzyme Ext1 in mouse brain results in severe defects in the formation of major commissural tracts and guidance errors of retinal axons at the optic chiasm.[22] In *Ext1* null mice, retinal axons from one eye misproject into the optic nerve of the contralateral eye. A similar misrouting of retinal axons at the optic chiasm also occurs in *Slit1/Slit2* double knockout mice,[23] but is absent in *Slit2* single mutant mice. Lowering the *Ext1* gene dosage in a *Slit2* null background resulted in severe retinal axon guidance errors, similar to *Ext1* null mutants and *Slit1/Slit2* double null mutants, showing a genetic interaction between Slit2 and HS.[22] In a genetic analysis of zebrafish mutants identified in a screen for retinal axon projection defects, Lee et al cloned the zebrafish Ext2 and Ext13 genes.[24] Loss of one of these HS-synthesizing enzyme encoding genes results in defects in the topographical sorting of retinal axons in the optic tract, but does not affect retinal axon pathfinding.[24] Upon loss of both Ext genes however, retinal axons display severe guidance defects in addition to optic tract sorting defects, which phenocopy the retinal pathfinding errors in *Robo2* mutant fish.[24,25] Together, these studies in mouse and zebrafish demonstrated an essential role for HS in Slit-mediated guidance of retinal axons at the optic chiasm in vivo.

Whether Slit proteins bind HS chains carried by a specific proteoglycan core protein could not be concluded from these studies. Recent work in *Drosophila* in which a HSPG core protein was deleted showed a critical, cell-autonomous role for the transmembrane HSPG syndecan in Slit-mediated axon guidance at the midline.[26,27] Loss of syndecan from axons resulted in midline guidance errors similar to those observed in *Slit* and *Robo* mutant flies.[26,27] Interestingly, deletion of syndecan in *Drosophila* also affected the extracellular distribution of Slit protein.[27] In *syndecan* mutants, Slit protein could still be detected on the surface of Slit-producing midline glial cells, but was no longer present on the surface of axons, suggesting that the HSPG syndecan regulates the extracellular distribution of Slit protein on axons. In line with this, the phenotype that results from loss of syndecan in *Drosophila* could be completely rescued by expression of syndecan in neurons, but not in midline cells, showing that syndecan functions in neurons.[26,27] The role of *Drosophila* syndecan in Slit-mediated midline guidance is in apparent contrast to the binding of mammalian Slit2 to glypican-1.[15] However, axonal expression of the membrane-attached HSPG glypican also rescued the syndecan mutant phenotype,[27] suggesting functional redundancy at the level of the HS-carrying core protein. A cell-autonomous role for syndecan in Slit-Robo signaling in midline axon guidance was also found in *C. elegans*.[28] The precise mechanism of HSPG function in modulating Slit-Robo signaling at the midline is still unclear. Although these results suggest a role in regulating the local distribution of Slit ligands, syndecan was also found to bind to both Slit and Robo,[27] suggesting that HSPGs could also act as essential coreceptors.

An additional level of complexity in the role of HSPGs in modulating Slit-mediated axon guidance is achieved by secondary modifications of HS by sulfotransferase and epimerase enzymes. The modifications of the HS chain by these enzymes are highly variable and result in an enormous variation in patterns of sulfation and epimerization, which are thought to play a role in the interaction of distinct proteins with HS.[29] The importance of specific HS sulfation patterns was demonstrated in the *Xenopus* visual system, where inhibition of HS sulfation or exogenous application of experimentally desulfated heparin caused retinal axon pathfinding defects similar to those resulting from enzymatic removal of HS.[13] Analyses of *C. elegans* mutants lacking one or more HS-modifying enzymes revealed marked axon guidance defects, showing that distinct HS modifications are required for specific guidance decisions in vivo. Some of these pathfinding errors could be linked to the Slit-Robo signaling pathway.[30] Slit-Robo-mediated guidance of specific classes of inter- and motor neurons was shown to depend on secondary HS modification by one epimerase and a sulfotransferase in some, but not all cases, indicating that Slit-mediated axon guidance requires HS modification in a cell type-specific manner.[28,30] Thus, the precise local composition of HS chains could critically influence axon guidance decisions at specific choice points. Taken together, genetic evidence from several model systems has shown an essential role for HSPGs in modulating Slit-Robo

Netrins and Heparan Sulfate Proteoglycans

Netrins were purified from embryonic brain membrane fractions using heparin columns, suggesting that netrins can bind HSPGs.[8] When expressed in heterologous cells, netrins associate with the cell surface.[31] Netrins contain a C-terminus enriched in basic residues, which could mediate its binding to negative charges on the cell surface including ECM components such as glycosaminoglycans.[8,32] In agreement with this, analysis of the protein distribution of netrins in embryonic and adult neural tissue has shown that netrin proteins are predominantly associated with cell surfaces and act at short-range. Netrin-1 protein in the embryonic mouse retina is associated with netrin-producing glial cells near the optic disc and acts locally at a short-range to guide retinal axons into the optic nerve.[33] *Drosophila* netrin proteins are detected on the cell surface of the midline glial cells that produce them and on the surface of commissural axons.[34] To directly test whether midline-derived netrins act at a short- or long-range in the guidance of midline crossing commissural axons in *Drosophila*, Brankatschk and Dickson employed a gene targeting strategy to tether the endogenous netrin B protein to the membrane of the netrin-secreting cells at the midline.[35] In the complete absence of netrins, commissural axons still reached the midline but failed to cross it. Anchoring netrin B to the cell surface abolished its long-range activity but preserved its short-range attractive function. In the absence of long-range guidance by netrin, commissural axons grew normally towards the midline and managed to cross it. These results suggest that midline-derived netrins act locally at short-range to facilitate midline crossing by commissural axons, rather than acting as long-range attractants.[35] A detailed immunohistochemical examination of netrin-1 distribution in the embryonic chick nervous system supports the idea that netrin-1 does not diffuse far from its source.[36] In the adult rat spinal cord, biochemical and immunohistochemical analyses showed that netrin-1 protein is not freely soluble but is associated with membranes or ECM.[37] Although the localization of netrins in tissue suggests an association with ECM constituents such as proteoglycans, the exact binding partner of netrins has not been identified.

Association with ECM components could play a role in localizing netrin protein in tissue, but has also been shown to regulate netrin function. The attractive turning of retinal growth cones towards a gradient of netrin-1 is converted to repulsion when the ECM molecule laminin-1 is present in the culture medium.[38] In agreement with this, a laminin-1 peptide mimetic caused retinal axon guidance errors at the optic nerve head, when applied to the developing retina in culture. Thus, depending on the context in which it is presented, netrin-1 can act as an attractant or a repellent, showing that ECM molecules can modulate the growth cone response to this secreted guidance cue.

Functional evidence that proteoglycans modulate netrin function in axon guidance in vitro or in vivo is currently lacking. Heparin has been implicated as part of the netrin receptor complex, either through direct binding to the netrin receptor DCC[39] or through indirect binding to DCC via netrin.[40] Nonetheless, HS does not seem to be absolutely required for netrin binding to its receptor, as competition experiments with excess heparin do not interfere with netrin binding to DCC.[40,41] Experiments in which netrin binding to DCC was analyzed following enzymatic removal of endogenous HS have not been reported to our knowledge. The above experiments suggest that netrin's interaction with heparin through its C-terminal domain mainly serves a role in protein distribution and is not required for its biological activity. However, striking phenotypic similarities between *Ext1* null mice and mice defective in netrin-DCC signaling have suggested a role for HS in modulating netrin-mediated axon guidance in vivo. As discussed in the previous paragraph, disruption of the HS-polymerizing enzyme Ext1 in the nervous system results in severe defects in mouse forebrain commissures.[22] In *Ext1* null mice, the corpus callosum, hippocampal commissure and anterior commissure are absent and the axons that normally give rise to these commissural tracts are misguided, resulting in the formation of Probst bundles.[22] Intriguingly, the absence of these commissures and the formation of Probst bundles also occur in *netrin* and *DCC* knockout mice.[42,43] The finding that commissural axon guidance defects occurring in *netrin/DCC* deficient mice are phenocopied in *Ext1* null mice, together with netrin's heparin binding properties,[8] suggests that HS could play a role in netrin-mediated guidance of forebrain commissural axons. Interestingly, this idea is further supported by the absence of hippocampal and anterior commissures resulting from disruption of the HS-modifying sulfotransferase Ndst1 in mice.[44] On the other hand, no evidence was found for a role of HS modification by three different HS modifying enzymes in netrin-mediated guidance in *C. elegans*.[30] Thus, it remains to be seen whether HS affects netrin function or that of other soluble guidance molecules in the wiring of forebrain commissural axons.

Secreted Semaphorins and Heparan Sulfate Proteoglycans

The secreted semaphorin Sema3A was purified from adult brain membrane fractions.[45] Secreted semaphorins contain a C-terminal basic region that could mediate their binding to negative charges on cell surface and ECM components.[45] Upon expression in neuroblastoma cells, recombinant Sema3A associates with the cell surface and can be removed from the cell surface with exogenous glycosaminoglycans or a glycosaminoglycan-degrading enzyme, suggesting that secreted semaphorins could bind to ECM components such as proteoglycans.[46] Detailed mapping studies of the protein distribution of secreted semaphorins in the nervous system are not available. In the human subiculum, Sema3A has been reported to show a punctate surface labeling of cell body and dendrites.[47] In the human cerebellum, Sema3A displays a punctate distribution around Purkinje cells, suggestive of secretion of Sema3A from Purkinje cell dendrites into the ECM or association with presynaptic terminals originating from other sources.[48] In rat Purkinje cells, the same antibody produced a different, intracellular staining pattern.[48] Sema3F displays a vesicular distribution in human cell lines and labels processes and nerve terminals in adult human brain.[49] Thus, secreted semaphorins appear to localize to cell surfaces, nerve terminals and intracellular organelles.

To date, there is no direct evidence that proteoglycans modulate axon guidance by secreted semaphorins, but there are some indications that suggest a role for proteoglycans in secreted semaphorin function. Neuropilin-1, an essential component of the Sema3A receptor complex,[50,51] can bind heparin.[52] Binding of vascular endothelial growth factor 165, a neuropilin-1 ligand that is structurally distinct from secreted semaphorins,[53] to neuropilin-1 and neuropilin-1-expressing cells is enhanced by addition of heparin.[52,54] A similar enhancing effect of heparin addition was observed on the growth cone collapsing activity of Sema3A, suggesting that association of Sema3A with HS could serve to potentiate Sema3A signaling.[46] However, HS removal of *Xenopus* retinal axons by heparinase treatment does not affect Sema3A-induced growth cone collapse.[21] Furthermore, Kantor et al observed no effect of heparinase treatment on alkaline phosphatase (AP)-tagged Sema3F binding to tissue sections or on Sema3F-mediated repulsion of axons originating from the habenula nucleus,[55] arguing against a role of HS in secreted semaphorin function in these neurons, at least in vitro. A clue for a role of HS in secreted semaphorin function in vivo may come from the developmental defects in commissural tract formation in the *Ext1* knockout mouse.[22] Secreted semaphorin signaling has been shown to be essential for the formation of forebrain commissural tracts. The anterior commissural tract is severely disorganized following loss of *Sema3F*, *Sema3B*, *neuropilin-2* and *Plexin A4*.[56-61] Furthermore, disruption of secreted semaphorin signaling through neuropilin-1 causes defects in corpus callosum formation[62] and loss of *Plexin A3* results in mistargeting of hippocampal commissural axons.[63] The phenotypic similarities in commissural tract malformation between mice defective in secreted semaphorin signaling and *Ext1* mice are intriguing and suggest a role for HSPGs in secreted semaphorin-mediated guidance of commissural axons.

Eph Receptors and Heparan Sulfate Proteoglycans

The transmembrane HSPG syndecan-2 plays a critical role in the development of dendritic spines. Syndecan-2 becomes localized to dendritic spines as hippocampal neurons mature in culture.[64] Interestingly, overexpression of syndecan-2 in young, immature neurons induces the formation of dendritic spines, an effect which is dependent on the PDZ domain binding site in the intracellular domain of syndecan-2.[64] The receptor tyrosine kinase EphB2 colocalizes with syndecan-2, coimmunoprecipitates with syndecan-2 and phosphorylates syndecan-2.[65] Furthermore, phosphorylation of syndecan-2 is required for clustering of syndecan-2 and syndecan-2-induced formation of dendritic spines. Conversely, inhibition of the kinase activity of EphB2 prevents syndecan-2 clustering and spine formation. Together, these studies suggest that activation of postsynaptic EphB2, possibly by presynaptic B-ephrins,[66,67] phosphorylates and clusters syndecan-2, which could subsequently recruit cytoplasmic effector proteins via its PDZ domain binding site. The signaling pathways initiated by these effector proteins could then lead to structural alterations in dendritic spines. One syndecan-2-binding protein is synbindin.[68] Immuno-electron microscopy showed that synbindin localizes to vesicles in the dendrite, suggesting that syndecan-2 may regulate the trafficking of vesicles at postsynaptic sites that eventually become dendritic spines.

The functional significance of Eph receptor-HSPG interactions in axon guidance in vitro or in vivo is at present uncertain. Similar to mice defective in secreted semaphorin and netrin signaling, Eph receptor mutant mice display forebrain commissure guidance defects[69,70] which partially phenocopy those of *Ext1* null mice.[22] In *C. elegans*, HS modification does not appear to play a role in ephrin-mediated guidance of specific interneuron axons, although a role for HS modification in ephrin-mediated axon guidance was not ruled out in this study.[30]

Semaphorin5A and Heparan Sulfate and Chondroitin Sulfate Proteoglycans

Compelling evidence for a role of proteoglycans as modulators of semaphorin function comes from a recent study by Kantor et al on the transmembrane semaphorin Sema5A.[55] Sema5A is expressed by neurons in the habenula nucleus (Hb) which give rise to the fasciculus retroflexus. In addition, Sema5A is expressed in prosomere 2, a subdivision of the embryonic forebrain, which is normally avoided by Hb axons (Fig. 1). Surprisingly, membrane preparations of HEK 293 cells expressing Sema5A act as an attractant for Hb axons, whereas Sema5A in prosomere 2 membrane preparations has a repellent influence on Hb axons.[55] Thus, Sema5A functions as a bifunctional guidance cue, exerting both attractive and repulsive effects on the same population of Hb axons. Sema5A contains two clusters of thrombospondin repeats, protein domains which are known to interact with GAGs,[71] suggesting that proteoglycans might interact with Sema5A. Indeed, heparinase treatment and pharmacological interference with HSPG synthesis in axons abolished the attractive effect of Sema5A, showing a cell-autonomous requirement for HSPGs in Hb axons for the attractive effect of Sema5A (Fig. 1). A candidate HSPG is syndecan-3, which is expressed on Hb axons and coimmunoprecipitates with Sema5A. The trajectory of Hb axons in *Sema5A* null mutant mice has not been determined as *Sema5A* null mutants die around embryonic day 12 in utero.[72] However, the fasciculus retroflexus is defasciculated in *Ext1* null mice,[55] indicating a requirement for HS in the guidance of Hb axons in vivo. Together, these data suggest that an HSPG expressed on the surface of Hb axons acts as an essential component of a receptor mediating the attractive effect of Sema5A.

Intriguingly, the repellent effect of Sema5A in prosomere 2 membranes on the same population of Hb axons was found to depend on a different GAG. In the presence of CSPGs, Sema5A converted from an attractive cue into an inhibitory cue for Hb axons.[55] Enzymatic removal of CS by chondroitinase treatment abolished the repulsive effect of prosomere 2 membranes on Hb axons. Therefore, CSPGs act as modulators of Sema5A function by switching this attractive guidance cue to an inhibitory cue. Taken together, this study demonstrates that the attractive effect of Sema5A on Hb axons depends on axonally expressed HSPGs, whereas

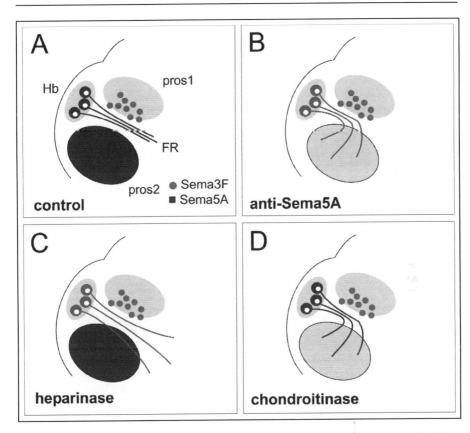

Figure 1. HSPGs and CSPGs regulate Sema5A function. A-D) Schematic lateral view of the embryonic (E13.5-15.5) rat diencephalon, based on (ref. 55). A) axons of neurons in the habenula nucleus (Hb) are funneled between prosomere 1 (pros1) and prosomere 2 (pros2) and form the fasciculus retroflexus (FR) in control embryos. Pros1 secretes the chemorepellent Sema3F (red dots).[56,120] Sema5A is expressed on Hb axons and in pros2 and interacts with endogenous binding partners in these locations, which can be visualized by incubating sections with an alkaline phosphatase (AP)-tagged Sema5A ectodomain (AP-Sema5A^{ecto}) fusion protein (indicated with blue color).[55] B) treatment of diencephalic explants with Sema5A function blocking antibodies (schematically indicated with the absence of blue coloring of the FR and pros2) results in the aberrant invasion of the normally avoided pros2 by Hb axons. C) treatment of diencephalic explants with heparinase reduces the binding of AP-Sema5A^{ecto} to the FR (absence of blue coloring), but not to pros2. In the absence of HS in Ext1 null mice, the FR is defasciculated.[55] In the presence of axonal HSPGs, Sema5A normally functions as an attractant to maintain the FR as a tightly fasciculated tract. D) treatment of diencephalic explants with chondroitinase reduces the binding of AP-Sema5A^{ecto} to pros2 (absence of blue coloring) but not to the FR, and results in the aberrant invasion of pros2 by Hb axons.[55] In the presence of CSPGs in the pros2 environment, Sema5A normally acts as a repellent to channel the FR between pros1 and pros2. A color version of this figure is available online at www.Eurekah.com.

the inhibitory effect of Sema5A on Hb axons depends on CSPGs in the prosomere 2 environment (Fig. 1). This shows that the composition of sulfated proteoglycans in the extracellular environment can critically modulate the growth cone response to this membrane-bound guidance cue, reminiscent of the laminin-induced conversion of attraction to repulsion of the secreted guidance cue netrin-1.[38] Thus, the specific local composition of the ECM can modulate axon guidance decisions to secreted and membrane-bound guidance cues to a great extent. The exact molecular mechanism by which CSPGs convert Sema5A's attractive function into an inhibitory function is at present poorly understood. CSPGs could act as a coreceptor and activate intracellular signaling molecules. Alternatively, CSPGs could promote interactions of Sema5A with a repulsive receptor or interfere with other receptor components that mediate the attractive effect of Sema5A. Several studies indicate that CSPGs can associate with guidance cues to modulate growth cone guidance, suggesting that modulation of axon guidance function by CSPGs may constitute a more general theme. The role of CSPGs in axon guidance will be discussed in the next paragraph.

Role of Chondroitin Sulfate Proteoglycans in Axon Guidance

Indications for a role of CSPGs in axon guidance come from in vitro and in vivo studies showing aberrant axon tract development following experimental manipulation of CS levels. Several studies have focused on the role of CSPGs in axon guidance in the visual system. In most cases, the CS-binding axon guidance cues mediating these effects have not been identified, with the exception of Sema5A-CSPG interactions in the guidance of Hb axons (discussed in the previous paragraph). Below, we will discuss the evidence pointing towards a role of CSPGs in regulating guidance of retinal, thalamic, motor and midbrain commissural axons, and speculate on the axon guidance cues involved.

Initial evidence for the involvement of CSPGs in axon guidance was provided with the demonstration that the treatment of retinal explant cultures with chondroitinase[73] or exogenous CS[74] caused retinal axons to grow into a region within the retina that would normally exert a repulsive influence on these axons.[75] The finding that both CS removal by chondroitinase treatment and addition of exogenous CS resulted in aberrant axon growth suggested that these treatments displaced repulsive guidance cues associated with CS. Experiments on retinal axon guidance in the *Xenopus* visual system in vivo showed that exogenous CS, though not chondroitinase treatment, caused retinal axons to defasciculate widely from their trajectory within the optic tract and to invade forebrain regions normally avoided by these axons (Fig. 2).[76] A similar effect has been found in chick embryos, where removal of endogenous CS by chondroitinase injection resulted in an anterior enlargement of the optic tract and aberrant invasion of the telencephalon.[77] Remarkably, the retinal axon guidance defects induced by CS treatment in *Xenopus* embryos were clearly distinct from the retinal axon pathfinding defects caused by heparin/HS treatment in the same experimental paradigm, which consisted of a characteristic 'bypass' of the tectum (Fig. 2).[12,13] This suggests that exogenous CS might displace a specific CS-binding factor in the *Xenopus* visual system. A candidate CS-binding cue is Sema3A, which is strongly expressed dorsally near the *Xenopus* optic tract (Fig. 2) and repels retinal axons.[78] In vitro, recombinant Sema3A can be removed from cell surfaces by exogenous CS or chondroitinase treatment.[46] Another candidate CS-binding cue is netrin-1, which is expressed dorsally to the optic tract and can repel retinal axons (Fig. 2).[79] Netrin-1 can bind to the small CSPG decorin (Dr. D. Litwack, personal communication and see ref. 80). Interestingly, blocking Sema5A function in mouse organotypic culture also causes retinal axons to stray out of the optic nerve.[81] The expression of Sema5A in the *Xenopus* visual system has not been reported however. In mouse brain slice preparations, chondroitinase treatment results in retinal axon guidance defects at the optic chiasm, including defasciculation, growth into inappropriate regions and misrouting into the contralateral optic stalk.[82] Ephrins,[83,84] slits[23,85] and netrins[86] have all been shown to regulate axon routing at the optic chiasm, but whether chondroitinase treatment affects localization or biological activity of any of these guidance cues at this important choicepoint is currently unknown. Besides affecting the localization of guidance cues, manipulation of CS could also modulate adhesive interactions along the optic tract.[87,88]

Outside the visual system, indications for a functional association of guidance cues with CSPGs come from a study on the adhesion and outgrowth of thalamic neurons.[89] When plated on mouse embryonic cortical slices, thalamic neurons normally avoid the cortical plate region

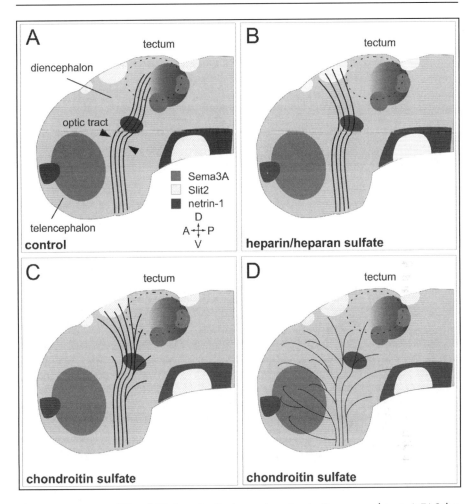

Figure 2. Exogenous HS and CS disrupt retinal axon targeting in *Xenopus* embryos. A-D) Schematic lateral view of stage 39-41 *Xenopus* brain, based on (refs. 12, 13, 21, 76, 78). A, projection of retinal axons in control embryos in relation to secreted axon guidance cue expression. The approximate regions of guidance cue expression are colored. The retinal axons (arrowheads) normally project through the optic tract into the optic tectum (indicated with dotted line), avoiding the telencephalic and diencephalic regions. Sema3A (red) is strongly expressed in a region in the telencephalon dorsal to the optic tract and in the posterior tectum in close proximity to retinal axons.[78] Slit2 (yellow) is expressed in the anterior and posterior margin of the tectum, as well as the dorsomedial border of the tectum (not shown) and in the diencephalon anterior to the optic tract.[21] Netrin-1 (blue) is expressed in a dense patch in the diencephalon and in a posterior to anterior gradient from the midbrain/hindbrain boundary into the optic tectum.[79] B, addition of exogenous heparin or HS severely disrupts retinal axon targeting. The axons take an aberrant dorsal turn in the diencephalon and bypass their target.[12,13] Exogenous HS might interfere with Slit2 function, which is dependent on HS in retinal axon guidance.[21] Another retinal axon guidance factor likely to be affected by HS treatment is FGF-2 (not shown).[12] C,D) various phenotypes following addition of exogenous CS. Exogenous CS severely disrupts retinal axon targeting. The axons are widely dispersed from their normal trajectory in the optic tract (C) and invade the normally avoided telencephalon and diencephalon (D).[76] Exogenous CS could interfere with netrin-1 or Sema3A function in retinal axon guidance. D, dorsal; V, ventral; A, anterior; P, posterior. A color version of this figure is available online at www.Eurekah.com.

and preferentially adhere to the subplate and intermediate zone. Interestingly, treatment with chondroitinase abolished these layer-specific differences, allowing attachment and neurite outgrowth of thalamic neurons on the formerly inhibitory cortical plate region. Competition experiments with exogenous CS mimicked the effects of chondroitinase treatment, strongly suggesting that CS-bound guidance cues mediated the layer-specific effects.[89] An interesting candidate for the CS-associated repellent activity in the cortical plate is Sema3A. Sema3A is strongly expressed in the embryonic cortical plate but not in the intermediate zone[90,91] and a repulsive gradient of Sema3A has been shown to be present in cortical slices.[92] Furthermore, Sema3A acts as a repellent for thalamic axons.[93] In *Sema3A* knockout mice however, the thalamocortical projection appears normal,[91,94] suggesting the presence of additional cues to guide thalamic axons to their cortical targets.

Chondroitinase treatment of zebrafish embryos resulted in the formation of abnormal sidebranches of ventral motor nerves.[95] This abnormal motor axon outgrowth could be partially reproduced by treatment with exogenous CS B, but not with CS C or heparinase, suggesting that differences in the sulfation pattern of CS sidechains might influence the binding of guidance cues. Following CS removal, some ventral motor nerve sidebranches extended into the posterior half of the somite, a region normally avoided by these axons. Interestingly, the ventral motor axon repellent Sema3A2, closely related to the mammalian Sema3A, is expressed in the posterior half of the somite,[96] suggesting that the manipulation of CS levels affected the localization of Sema3A2 in this region.

Following exposure to exogenous CS, axons in the postoptic commissural tract in *Xenopus* embryos fail to make a turn into the midbrain ventral commissure, and instead continue to grow longitudinally or instead make an aberrant dorsal turn.[97] Interfering with the function of the netrin receptor DCC, which is expressed on these axons, caused a similar phenotype, suggesting that netrin-1 is involved in the guidance of forebrain axons in the ventral commissure.[98] The addition of exogenous CS could displace netrin-1 from CSPGs, thereby removing an attractive cue that normally directs these axons into the ventral commissure and exposing a repulsive cue, possibly Sema3A,[99] which causes forebrain axons to turn dorsally. Taken together, several studies have indicated a role for CSPGs in modulating axon guidance in mouse, *Xenopus* and zebrafish embryos. Although direct evidence for the CS-binding guidance cues involved is currently lacking, these studies hint at the intriguing possibility that CSPGs modulate secreted semaphorin and netrin function.

Proteoglycans and Guidance Molecules in the Regeneration of Adult Neurons

Proteoglycans and several axon guidance molecules continue to be expressed in the mature nervous system. The failure of regeneration of adult CNS neurons is believed to be the result of multiple intrinsic and extrinsic causes. A major contributing factor to regenerative failure is the extracellular CNS environment. As described in the preceding pages a number of axon guidance molecules collaborate with proteoglycans to form molecular barriers that are important for the correct and stereotypical guidance of developing axons. Do these interactions also occur in the mature and injured nervous system? And are these interactions part of a biological scenario that is prohibitive to synapse formation during adulthood and axonal regrowth following neural injury?

CSPG expression increases quickly after a spinal cord injury and occurs in meningeal cells, reactive astrocytes and in cells of the oligodendrocyte lineage. Astrocytes and oligodendrocytes express neurocan, phosphacan and brevican. Versican and NG2 are synthesized by oligodendrocyte lineage cells associated with the scar. Also the GAG contents of some of the CSPGs formed in the scar increases.[100,101] Evidence for an inhibitory role of the CSPG and HSPG GAG chains has been obtained in vitro and in vivo. In a culture system of the glial scar the inhibitory effects of the scar towards neurite outgrowth could be overcome by treatment with chondroitinase ABC (chABC).[102,103] More recently this enzyme was applied in vivo in several regeneration paradigms. Impressive regeneration of the lesioned adult rat nigrostriatal pathway occurred after treatment of this brain lesion with chABC.[104] Infusion of chABC in the spinal cord following a lesion of dorsal column axons provoked regeneration of sensory fibers.[105] In this study enhanced fiber regeneration was accompanied with upregulation of growth-associated protein-43 in the regenerating neurons and a certain degree of functional improvement. ChABC treatment has now been combined with a number of other regeneration promoting treatments usually resulting in a more potent regenerative response. For instance, Schwann cell cables that serve as permissive conduits for regenerating spinal cord axons are normally sealed off at their distal end by a CSPG rich scar. This problem appears to be solved to some extent by infusing chABC distally to implanted Schwann cell cables, because this resulted in significant regeneration of fibers from the transplant into the distal spinal cord.[106] Although much more is known about the role of CSPGs than HSPGs in neuroregeneration, HSPGs have also been implicated in the inhibition of the regeneration of injured peripheral axons. Treatment of peripheral nerve grafts with heparinase I prior to transplantation into a transected peripheral nerve induced a twofold increase in the number of regenerating axons in these grafts.[107] Whether the enhanced regenerative axonal growth is the result of removal of (a) protein(s) that bind to the HS-GAG chains, to the removal of HS-GAG chains as such, or another mechanism is currently not known.

Interestingly, infusion of chABC into the visual cortex reactivates ocular dominance plasticity in the adult visual cortex.[108] Normally, neurons in the visual cortex loose their anatomical plasticity around the end of the first month after birth. The stabilization of synapses in the visual cortex coincides with the formation of perineuronal nets around cortical neurons. These nets are composed of ECM molecules, including CSPGs, which are denuded from their GAG chains by chABC. This apparently frees these neurons from an inhibitory activity resulting in prolonged plasticity.[108] Thus, removal of the GAG chains of CSPGs is a strategy to promote injured nerve fibers to regenerate and affects the anatomical plasticity of intact, stable neuronal circuits in the adult. While the beneficial effects of chABC treatment have now been shown in various paradigms, the molecular mechanism by which CSPGs prevent neurite outgrowth and synaptic plasticity are poorly understood. To our knowledge, no neuronal cell surface receptors that interact directly with CSPGs have been identified. Thus, the signaling events that mediate the effects of CSPG remain largely elusive.

Members of almost all families of guidance molecules are expressed by cells of the neural scar or in oligodendrocytes. Secreted semaphorins occur in the meningeal fibroblasts that invade the center of the lesion within the first weeks.[109,110] Reactive astrocytes express Ephrin-B2[111] and Slit2.[19] Oligodendrocytes around the scar express the transmembrane semaphorins Sema4D,[112] Sema5A[113] and EphrinB3.[114]

The expression of CSPG and axonal guidance cues following neural injury in the CNS have mostly been studied separately. The expression patterns of Sema3A in relation to other proteins implicated in neurite outgrowth inhibition, the CSPGs, tenascin-C and myelin-derived inhibitors has been studied in some detail in transection lesions of the dorsal column and in relation to sensory fiber regeneration.[115] CSPG and tenascin-C expression overlapped with Sema3A in the meninges and in the dorsolateral cap of scar tissue that is mostly comprised of meningeal fibroblasts. The area of expression of tenascin-C and CSPG extended deeper into the ventral aspect of the lesion where no Sema3A positive cells were present. Conditioning lesions enabled injured ascending sensory axons to regrow across areas of strong CSPG and tenascin-C expression, while areas containing both Sema3A as well as CSPGs in the dorso-lateral portion of the scar were avoided by regenerating sensory fibers.[115] Thus, conditioning lesions that enhance the growth state of ascending sensory neurites[115,116] promoted nerve sprouting into areas of CSPG expression but failed to induce outgrowth into CSPG-Sema3A positive areas of the neural scar.

The growth inhibitory properties of meningeal fibroblasts are associated with the cell-surface rather than with soluble factors. Conditioned media from cultured meningeal fibroblasts do not block neurite outgrowth[117] and do not induce growth cone collapse.[118] Cocultured astrocytes and meningeal fibroblasts cluster in separate patches, thereby defining distinct cellular territories for postnatal DRG neurons. These neurons do not cross the interface between astrocytes and meningeal fibroblasts but their axons can cross to astrocytes when seeded on fibroblasts.[119] This is another indication that inhibitory factors are localized on the cell surface or ECM and that the meningeal cell is an important source of these inhibitors. In this coculture system perturbation of semaphorin signaling by a neuropilin-2 antibody partially reversed the repellent effect of the meningeal boundary.[119] Membrane extracts from cultured adult and neonatal meningeal fibroblast induce collapse of embryonic DRG growth cones, which can be blocked by neuropilin-1 antibodies and is absent in membrane extracts from Sema3A-deficient meningeal fibroblasts.[118] Meningeal fibroblasts from Sema3A knockout mice are a more permissive substrate than cells from wild-type littermates. Thus Sema3A is a major neurite growth-inhibitory factor in meningeal fibroblasts and appears to be presented as a substrate molecule associated with the cell membrane or the ECM.[118] As described above, removing the GAG chains with chABC or competing for binding with soluble CS releases recombinant Sema3A into the medium, indicating that Sema3A can bind to CS chains.[46] As previously mentioned, the mechanisms by which CSPGs inhibit neurite outgrowth are poorly understood. One explanation for the failure of regenerating axons to grow into CSPG-Sema3A positive scar tissue may be that repulsive proteins that bind to the GAG-chains of CSPG, e.g., Sema3A, form inhibitory molecular complexes with each other. Work in the developing nervous system provides evidence for the presence of differentially localized CS-binding molecules, conferring specific inhibitory or stimulatory activity.[89] Interactions with HSPGs and CSPGs can convert the transmembrane Sema5A from an attractive into an inhibitory axon guidance cue.[55] Sema5A is one factor produced by oligodendrocytes that may inhibit regeneration in the optic nerve[113] and perhaps elsewhere in the CNS. Because CSPGs are upregulated after a lesion this may result in a relative enrichment of highly repulsive Sema5A-CSPG complexes in the injured CNS. Whether CSPG-semaphorin interaction indeed occurs in the scar and how this affects the inhibitory activity of these molecules on regenerating axons is a fascinating question that has to be addressed in future studies on the inhibitory mechanisms active in the neural scar.

Concluding Remarks

Over the past few years, developmental studies have begun to reveal an essential role for proteoglycans in regulating key signaling pathways in axon guidance. The importance of HSPGs as modulators of axon guidance cue function is best illustrated for the Slit family of axon guidance cues, where a crucial function of HSPGs is conserved from worm to mouse. Besides Slits, HSPGs in the form of syndecans also function in EphB2 and Sema5A signaling, suggesting a common theme in modulating axon guidance cue function. Recent work has also indicated the importance of CSPGs in modifying the guidance function of Sema5A. In addition, the axon guidance defects that arise from experimental manipulation of HS and CS suggest that other soluble guidance cues such as netrins and secreted semaphorins could also functionally associate with heparan and chondroitin sulfate proteoglycans. Taken together, it is becoming increasingly clear that different molecules in the ECM can critically modulate axon guidance decisions. Yet exactly how proteoglycans modulate axon guidance cue function is still poorly understood. In the case of Slit proteins, HSPGs might function both in regulating the extracellular localization of Slit ligands as well as in acting as an essential coreceptor in a Slit-Robo signaling complex. Whether proteoglycans have similar functions in regulating signaling by other guidance molecules is largely undetermined. So far, most studies have focused on the importance of GAG chains in modulating guidance cue function. Given the large diversity in HSPG and CSPG core proteins in the nervous system, it is likely that additional roles for the core proteins will be discovered as well. Investigating the axon guidance phenotypes following genetic disruption of genes encoding core proteins and GAG chain synthesizing and modifying enzymes, as well as determining the protein distribution of axon guidance cues following manipulation of HS/CS will further contribute to our understanding of the role of proteoglycans in modulating guidance cue function. Furthermore, given the importance of proteoglycans in impeding the regenerative response of adult injured axons and the presence of axon guidance cues such as secreted semaphorins in the injured spinal cord, it will be of great interest to determine whether proteoglycans continue to function as modulators of axon guidance cue function in the adult nervous system as well.

References

1. Tessier-Lavigne M, Goodman CS. The molecular biology of axon guidance. Science 1996; 274(5290):1123-33.
2. Dickson BJ. Molecular mechanisms of axon guidance. Science 2002; 298(5600):1959-64.
3. Bandtlow CE, Zimmermann DR. Proteoglycans in the developing brain: New conceptual insights for old proteins. Physiol Rev 2000; 80(4):1267-90.
4. Bernfield M, Gotte M, Park PW et al. Functions of cell surface heparan sulfate proteoglycans. Annu Rev Biochem 1999; 68:729-77.
5. Baeg GH, Perrimon N. Functional binding of secreted molecules to heparan sulfate proteoglycans in Drosophila. Curr Opin Cell Biol 2000; 12(5):575-80.
6. Lin X. Functions of heparan sulfate proteoglycans in cell signaling during development. Development 2004; 131(24):6009-21.
7. Sugahara K, Mikami T, Uyama T et al. Recent advances in the structural biology of chondroitin sulfate and dermatan sulfate. Curr Opin Struct Biol 2003; 13(5):612-20.
8. Serafini T, Kennedy TE, Galko MJ et al. The netrins define a family of axon outgrowth-promoting proteins homologous to C. elegans UNC-6. Cell 1994; 78(3):409-24.
9. Wang KH, Brose K, Arnott D et al. Biochemical purification of a mammalian slit protein as a positive regulator of sensory axon elongation and branching. Cell 1999; 96(6):771-84.
10. Raper JA, Kapfhammer JP. The enrichment of a neuronal growth cone collapsing activity from embryonic chick brain. Neuron 1990; 4(1):21-9.
11. Wang L, Denburg JL. A role for proteoglycans in the guidance of a subset of pioneer axons in cultured embryos of the cockroach. Neuron 1992; 8(4):701-14.
12. Walz A, McFarlane S, Brickman YG et al. Essential role of heparan sulfates in axon navigation and targeting in the developing visual system. Development 1997; 124(12):2421-30.
13. Irie A, Yates EA, Turnbull JE et al. Specific heparan sulfate structures involved in retinal axon targeting. Development 2002; 129(1):61-70.
14. Brose K, Bland KS, Wang KH et al. Slit proteins bind Robo receptors and have an evolutionarily conserved role in repulsive axon guidance. Cell 1999; 96(6):795-806.
15. Liang Y, Annan RS, Carr SA et al. Mammalian homologues of the Drosophila slit protein are ligands of the heparan sulfate proteoglycan glypican-1 in brain. J Biol Chem 1999; 274(25):17885-92.
16. Ronca F, Andersen JS, Paech V et al. Characterization of Slit protein interactions with glypican-1. J Biol Chem 2001; 276(31):29141-7.
17. Kidd T, Bland KS, Goodman CS. Slit is the midline repellent for the robo receptor in Drosophila. Cell 1999; 96(6):785-94.
18. Rothberg JM, Jacobs JR, Goodman CS et al. Slit: An extracellular protein necessary for development of midline glia and commissural axon pathways contains both EGF and LRR domains. Genes Dev 1990; 4(12A):2169-87.
19. Hagino S, Iseki K, Mori T et al. Slit and glypican-1 mRNAs are coexpressed in the reactive astrocytes of the injured adult brain. Glia 2003; 42(2):130-8.
20. Hu H. Cell-surface heparan sulfate is involved in the repulsive guidance activities of Slit2 protein. Nat Neurosci 2001; 4(7):695-701.
21. Piper M, Anderson R, Dwivedy A et al. Signaling mechanisms underlying slit2-induced collapse of Xenopus retinal growth cones. Neuron 2006; 49(2):215-28.
22. Inatani M, Irie F, Plump AS et al. Mammalian brain morphogenesis and midline axon guidance require heparan sulfate. Science 2003; 302(5647):1044-6.
23. Plump AS, Erskine L, Sabatier C et al. Slit1 and Slit2 cooperate to prevent premature midline crossing of retinal axons in the mouse visual system. Neuron 2002; 33(2):219-32.
24. Lee JS, von der Hardt S, Rusch MA et al. Axon sorting in the optic tract requires HSPG synthesis by ext2 (dackel) and extl3 (boxer). Neuron 2004; 44(6):947-60.

25. Fricke C, Lee JS, Geiger-Rudolph S et al. astray, a zebrafish roundabout homolog required for retinal axon guidance. Science 2001; 292(5516):507-10.
26. Steigemann P, Molitor A, Fellert S et al. Heparan sulfate proteoglycan syndecan promotes axonal and myotube guidance by slit/robo signaling. Curr Biol 2004; 14(3):225-30.
27. Johnson KG, Ghose A, Epstein E et al. Axonal heparan sulfate proteoglycans regulate the distribution and efficiency of the repellent slit during midline axon guidance. Curr Biol 2004; 14(6):499-504.
28. Rhiner C, Gysi S, Frohli E et al. Syndecan regulates cell migration and axon guidance in C. elegans. Development 2005; 132(20):4621-33.
29. Turnbull J, Powell A, Guimond S. Heparan sulfate: Decoding a dynamic multifunctional cell regulator. Trends Cell Biol 2001; 11(2):75-82.
30. Bulow HE, Hobert O. Differential sulfations and epimerization define heparan sulfate specificity in nervous system development. Neuron 2004; 41(5):723-36.
31. Kennedy TE, Serafini T, de la Torre JR et al. Netrins are diffusible chemotropic factors for commissural axons in the embryonic spinal cord. Cell 1994; 78(3):425-35.
32. Kappler J, Franken S, Junghans U et al. Glycosaminoglycan-binding properties and secondary structure of the C-terminus of netrin-1. Biochem Biophys Res Commun 2000; 271(2):287-91.
33. Deiner MS, Kennedy TE, Fazeli A et al. Netrin-1 and DCC mediate axon guidance locally at the optic disc: Loss of function leads to optic nerve hypoplasia. Neuron 1997; 19(3):575-89.
34. Harris R, Sabatelli LM, Seeger MA. Guidance cues at the Drosophila CNS midline: Identification and characterization of two Drosophila Netrin/UNC-6 homologs. Neuron 1996; 17(2):217-28.
35. Brankatschk M, Dickson BJ. Netrins guide Drosophila commissural axons at short range. Nat Neurosci 2006; 9(2):188-94.
36. MacLennan AJ, McLaurin DL, Marks L et al. Immunohistochemical localization of netrin-1 in the embryonic chick nervous system. J Neurosci 1997; 17(14):5466-79.
37. Manitt C, Colicos MA, Thompson KM et al. Widespread expression of netrin-1 by neurons and oligodendrocytes in the adult mammalian spinal cord. J Neurosci 2001; 21(11):3911-22.
38. Hopker VH, Shewan D, Tessier-Lavigne M et al. Growth-cone attraction to netrin-1 is converted to repulsion by laminin-1. Nature 1999; 401(6748):69-73.
39. Bennett KL, Bradshaw J, Youngman T et al. Deleted in colorectal carcinoma (DCC) binds heparin via its fifth fibronectin type III domain. J Biol Chem 1997; 272(43):26940-6.
40. Geisbrecht BV, Dowd KA, Barfield RW et al. Netrin binds discrete subdomains of DCC and UNC5 and mediates interactions between DCC and heparin. J Biol Chem 2003; 278(35):32561-8.
41. Keino-Masu K, Masu M, Hinck L et al. Deleted in Colorectal Cancer (DCC) encodes a netrin receptor. Cell 1996; 87(2):175-85.
42. Fazeli A, Dickinson SL, Hermiston ML et al. Phenotype of mice lacking functional Deleted in colorectal cancer (Dcc) gene. Nature 1997; 386(6627):796-804.
43. Serafini T, Colamarino SA, Leonardo ED et al. Netrin-1 is required for commissural axon guidance in the developing vertebrate nervous system. Cell 1996; 87(6):1001-14.
44. Grobe K, Inatani M, Pallerla SR et al. Cerebral hypoplasia and craniofacial defects in mice lacking heparan sulfate Ndst1 gene function. Development 2005; 132(16):3777-86.
45. Luo Y, Raible D, Raper JA. Collapsin: A protein in brain that induces the collapse and paralysis of neuronal growth cones. Cell 1993; 75(2):217-27.
46. De Wit J, De Winter F, Klooster J et al. Semaphorin 3A displays a punctate distribution on the surface of neuronal cells and interacts with proteoglycans in the extracellular matrix. Mol Cell Neurosci 2005; 29(1):40-55.
47. Good PF, Alapat D, Hsu A et al. A role for semaphorin 3A signaling in the degeneration of hippocampal neurons during Alzheimer's disease. J Neurochem 2004; 91(3):716-36.
48. Eastwood SL, Law AJ, Everall IP et al. The axonal chemorepellant semaphorin 3A is increased in the cerebellum in schizophrenia and may contribute to its synaptic pathology. Mol Psychiatry 2003; 8(2):148-55.
49. Hirsch E, Hu LJ, Prigent A et al. Distribution of semaphorin IV in adult human brain. Brain Res 1999; 823(1-2):67-79.
50. Kolodkin AL, Levengood DV, Rowe EG et al. Neuropilin is a semaphorin III receptor. Cell 1997; 90(4):753-62.
51. He Z, Tessier-Lavigne M. Neuropilin is a receptor for the axonal chemorepellent Semaphorin III. Cell 1997; 90(4):739-51.
52. Fuh G, Garcia KC, de Vos AM. The interaction of neuropilin-1 with vascular endothelial growth factor and its receptor flt-1. J Biol Chem 2000; 275(35):26690-5.
53. Soker S, Takashima S, Miao HQ et al. Neuropilin-1 is expressed by endothelial and tumor cells as an isoform-specific receptor for vascular endothelial growth factor. Cell 1998; 92(6):735-45.

54. Soker S, Fidder H, Neufeld G et al. Characterization of novel vascular endothelial growth factor (VEGF) receptors on tumor cells that bind VEGF165 via its exon 7-encoded domain. J Biol Chem 1996; 271(10):5761-7.
55. Kantor DB, Chivatakarn O, Peer KL et al. Semaphorin 5A is a bifunctional axon guidance cue regulated by heparan and chondroitin sulfate proteoglycans. Neuron 2004; 44(6):961-75.
56. Sahay A, Molliver ME, Ginty DD et al. Semaphorin 3F is critical for development of limbic system circuitry and is required in neurons for selective CNS axon guidance events. J Neurosci 2003; 23(17):6671-80.
57. Yaron A, Huang PH, Cheng HJ et al. Differential requirement for Plexin-A3 and -A4 in mediating responses of sensory and sympathetic neurons to distinct class 3 Semaphorins. Neuron 2005; 45(4):513-23.
58. Chen H, Bagri A, Zupicich JA et al. Neuropilin-2 regulates the development of selective cranial and sensory nerves and hippocampal mossy fiber projections. Neuron 2000; 25(1):43-56.
59. Giger RJ, Cloutier JF, Sahay A et al. Neuropilin-2 is required in vivo for selective axon guidance responses to secreted semaphorins. Neuron 2000; 25(1):29-41.
60. Julien F, Bechara A, Fiore R et al. Dual functional activity of Semaphorin 3B is required for positioning the anterior commissure. Neuron 2005; 48(1):63-75.
61. Suto F, Ito K, Uemura M et al. Plexin-a4 mediates axon-repulsive activities of both secreted and transmembrane semaphorins and plays roles in nerve fiber guidance. J Neurosci 2005; 25(14):3628-37.
62. Gu C, Rodriguez ER, Reimert DV et al. Neuropilin-1 conveys semaphorin and VEGF signaling during neural and cardiovascular development. Dev Cell 2003; 5(1):45-57.
63. Cheng HJ, Bagri A, Yaron A et al. Plexin-A3 mediates semaphorin signaling and regulates the development of hippocampal axonal projections. Neuron 2001; 32(2):249-63.
64. Ethell IM, Yamaguchi Y. Cell surface heparan sulfate proteoglycan syndecan-2 induces the maturation of dendritic spines in rat hippocampal neurons. J Cell Biol 1999; 144(3):575-86.
65. Ethell IM, Irie F, Kalo MS et al. EphB/syndecan-2 signaling in dendritic spine morphogenesis. Neuron 2001; 31(6):1001-13.
66. Henkemeyer M, Itkis OS, Ngo M et al. Multiple EphB receptor tyrosine kinases shape dendritic spines in the hippocampus. J Cell Biol 2003; 163(6):1313-26.
67. Penzes P, Beeser A, Chernoff J et al. Rapid induction of dendritic spine morphogenesis by trans-synaptic EphrinB-EphB receptor activation of the Rho-GEF Kalirin. Neuron 2003; 37(2):263-74.
68. Ethell IM, Hagihara K, Miura Y et al. Synbindin, A novel syndecan-2-binding protein in neuronal dendritic spines. J Cell Biol 2000; 151(1):53-68.
69. Henkemeyer M, Orioli D, Henderson JT et al. Nuk controls pathfinding of commissural axons in the mammalian central nervous system. Cell 1996; 86(1):35-46.
70. Orioli D, Henkemeyer M, Lemke G et al. Sek4 and Nuk receptors cooperate in guidance of commissural axons and in palate formation. EMBO J 1996; 15(22):6035-49.
71. Adams JC, Tucker RP. The thrombospondin type 1 repeat (TSR) superfamily: Diverse proteins with related roles in neuronal development. Dev Dyn 2000; 218(2):280-99.
72. Fiore R, Rahim B, Christoffels VM et al. Inactivation of the Sema5a gene results in embryonic lethality and defective remodeling of the cranial vascular system. Mol Cell Biol 2005; 25(6):2310-9.
73. Brittis PA, Canning DR, Silver J. Chondroitin sulfate as a regulator of neuronal patterning in the retina. Science 1992; 255(5045):733-6.
74. Brittis PA, Silver J. Exogenous glycosaminoglycans induce complete inversion of retinal ganglion cell bodies and their axons within the retinal neuroepithelium. Proc Natl Acad Sci USA 1994; 91(16):7539-42.
75. Snow DM, Watanabe M, Letourneau PC et al. A chondroitin sulfate proteoglycan may influence the direction of retinal ganglion cell outgrowth. Development 1991; 113(4):1473-85.
76. Walz A, Anderson RB, Irie A et al. Chondroitin sulfate disrupts axon pathfinding in the optic tract and alters growth cone dynamics. J Neurobiol 2002; 53(3):330-42.
77. Ichijo H, Kawabata I. Roles of the telencephalic cells and their chondroitin sulfate proteoglycans in delimiting an anterior border of the retinal pathway. J Neurosci 2001; 21(23):9304-14.
78. Campbell DS, Regan AG, Lopez JS et al. Semaphorin 3A elicits stage-dependent collapse, turning, and branching in Xenopus retinal growth cones. J Neurosci 2001; 21(21):8538-47.
79. Shewan D, Dwivedy A, Anderson R et al. Age-related changes underlie switch in netrin-1 responsiveness as growth cones advance along visual pathway. Nat Neurosci 2002; 5(10):955-62.
80. Litwack ED, Galko MJ, Danielson K et al. Decorin in the developing mouse nervous system; expression in the floorplate, and binding to netrin-1. Soc Neurosci Abstract 1995; 21:1022.

81. Oster SF, Bodeker MO, He F et al. Invariant Sema5A inhibition serves an ensheathing function during optic nerve development. Development 2003; 130(4):775-84.
82. Chung KY, Taylor JS, Shum DK et al. Axon routing at the optic chiasm after enzymatic removal of chondroitin sulfate in mouse embryos. Development 2000; 127(12):2673-83.
83. Nakagawa S, Brennan C, Johnson KG et al. Ephrin-B regulates the Ipsilateral routing of retinal axons at the optic chiasm. Neuron 2000; 25(3):599-610.
84. Williams SE, Mann F, Erskine L et al. Ephrin-B2 and EphB1 mediate retinal axon divergence at the optic chiasm. Neuron 2003; 39(6):919-35.
85. Niclou SP, Jia L, Raper JA. Slit2 is a repellent for retinal ganglion cell axons. J Neurosci 2000; 20(13):4962-74.
86. Deiner MS, Sretavan DW. Altered midline axon pathways and ectopic neurons in the developing hypothalamus of netrin-1- and DCC-deficient mice. J Neurosci 1999; 19(22):9900-12.
87. Friedlander DR, Milev P, Karthikeyan L et al. The neuronal chondroitin sulfate proteoglycan neurocan binds to the neural cell adhesion molecules Ng-CAM/L1/NILE and N-CAM, and inhibits neuronal adhesion and neurite outgrowth. J Cell Biol 1994; 125(3):669-80.
88. Grumet M, Flaccus A, Margolis RU. Functional characterization of chondroitin sulfate proteoglycans of brain: Interactions with neurons and neural cell adhesion molecules. J Cell Biol 1993; 120(3):815-24.
89. Emerling DE, Lander AD. Inhibitors and promoters of thalamic neuron adhesion and outgrowth in embryonic neocortex: Functional association with chondroitin sulfate. Neuron 1996; 17(6):1089-100.
90. Giger RJ, Wolfer DP, De Wit GM et al. Anatomy of rat semaphorin III/collapsin-1 mRNA expression and relationship to developing nerve tracts during neuroembryogenesis. J Comp Neurol 1996; 375(3):378-92.
91. Catalano SM, Messersmith EK, Goodman CS et al. Many major CNS axon projections develop normally in the absence of semaphorin III. Mol Cell Neurosci 1998; 11(4):173-82.
92. Polleux F, Giger RJ, Ginty DD et al. Patterning of cortical efferent projections by semaphorin-neuropilin interactions. Science 1998; 282(5395):1904-6.
93. Bagnard D, Chounlamountri N, Puschel AW et al. Axonal surface molecules act in combination with semaphorin 3a during the establishment of corticothalamic projections. Cereb Cortex 2001; 11(3):278-85.
94. Ulupinar E, Datwani A, Behar O et al. Role of semaphorin III in the developing rodent trigeminal system. Mol Cell Neurosci 1999; 13(4):281-92.
95. Bernhardt RR, Schachner M. Chondroitin sulfates affect the formation of the segmental motor nerves in zebrafish embryos. Dev Biol 2000; 221(1):206-19.
96. Roos M, Schachner M, Bernhardt RR. Zebrafish semaphorin Z1b inhibits growing motor axons in vivo. Mech Dev 1999; 87(1-2):103-17.
97. Anderson RB, Walz A, Holt CE et al. Chondroitin sulfates modulate axon guidance in embryonic Xenopus brain. Dev Biol 1998; 202(2):235-43.
98. Anderson RB, Cooper HM, Jackson SC et al. DCC plays a role in navigation of forebrain axons across the ventral midbrain commissure in embryonic xenopus. Dev Biol 2000; 217(2):244-53.
99. Anderson RB, Jackson SC, Fujisawa H et al. Expression and putative role of neuropilin-1 in the early scaffold of axon tracts in embryonic Xenopus brain. Dev Dyn 2000; 219(1):102-8.
100. Silver J, Miller JH. Regeneration beyond the glial scar. Nat Rev Neurosci 2004; 5(2):146-56.
101. Properzi F, Fawcett JW. Proteoglycans and brain repair. News Physiol Sci 2004; 19:33-8.
102. McKeon RJ, Hoke A, Silver J. Injury-induced proteoglycans inhibit the potential for laminin-mediated axon growth on astrocytic scars. Exp Neurol 1995; 136(1):32-43.
103. Rudge JS, Smith GM, Silver J. An in vitro model of wound healing in the CNS: Analysis of cell reaction and interaction at different ages. Exp Neurol 1989; 103(1):1-16.
104. Moon LD, Asher RA, Rhodes KE et al. Regeneration of CNS axons back to their target following treatment of adult rat brain with chondroitinase ABC. Nat Neurosci 2001; 4(5):465-6.
105. Bradbury EJ, Moon LD, Popat RJ et al. Chondroitinase ABC promotes functional recovery after spinal cord injury. Nature 2002; 416(6881):636-40.
106. Chau CH, Shum DK, Li H et al. Chondroitinase ABC enhances axonal regrowth through Schwann cell-seeded guidance channels after spinal cord injury. Faseb J 2004; 18(1):194-6.
107. Groves ML, McKeon R, Werner E et al. Axon regeneration in peripheral nerves is enhanced by proteoglycan degradation. Exp Neurol 2005; 195(2):278-92.
108. Pizzorusso T, Medini P, Berardi N et al. Reactivation of ocular dominance plasticity in the adult visual cortex. Science 2002; 298(5596):1248-51.
109. De Winter F, Oudega M, Lankhorst AJ et al. Injury-induced class 3 semaphorin expression in the rat spinal cord. Exp Neurol 2002; 175(1):61-75.

110. Pasterkamp RJ, Giger RJ, Ruitenberg MJ et al. Expression of the gene encoding the chemorepellent semaphorin III is induced in the fibroblast component of neural scar tissue formed following injuries of adult but not neonatal CNS. Mol Cell Neurosci 1999; 13(2):143-66.
111. Bundesen LQ, Scheel TA, Bregman BS et al. Ephrin-B2 and EphB2 regulation of astrocyte-meningeal fibroblast interactions in response to spinal cord lesions in adult rats. J Neurosci 2003; 23(21):7789-800.
112. Moreau-Fauvarque C, Kumanogoh A, Camand E et al. The transmembrane semaphorin Sema4D/ CD100, an inhibitor of axonal growth, is expressed on oligodendrocytes and upregulated after CNS lesion. J Neurosci 2003; 23(27):9229-39.
113. Goldberg JL, Vargas ME, Wang JT et al. An oligodendrocyte lineage-specific semaphorin, Sema5A, inhibits axon growth by retinal ganglion cells. J Neurosci 2004; 24(21):4989-99.
114. Benson MD, Romero MI, Lush ME et al. Ephrin-B3 is a myelin-based inhibitor of neurite outgrowth. Proc Natl Acad Sci USA 2005; 102(30):10694-9.
115. Pasterkamp RJ, Anderson PN, Verhaagen J. Peripheral nerve injury fails to induce growth of lesioned ascending dorsal column axons into spinal cord scar tissue expressing the axon repellent Semaphorin3A. Eur J Neurosci 2001; 13(3):457-71.
116. Neumann S, Woolf CJ. Regeneration of dorsal column fibers into and beyond the lesion site following adult spinal cord injury. Neuron 1999; 23(1):83-91.
117. Noble M, Fok-Seang J, Cohen J. Glia are a unique substrate for the in vitro growth of central nervous system neurons. J Neurosci 1984; 4(7):1892-903.
118. Niclou SP, Franssen EH, Ehlert EM et al. Meningeal cell-derived semaphorin 3A inhibits neurite outgrowth. Mol Cell Neurosci 2003; 24(4):902-12.
119. Shearer MC, Niclou SP, Brown D et al. The astrocyte/meningeal cell interface is a barrier to neurite outgrowth which can be overcome by manipulation of inhibitory molecules or axonal signaling pathways. Mol Cell Neurosci 2003; 24(4):913-25.
120. Funato H, Saito-Nakazato Y, Takahashi H. Axonal growth from the habenular nucleus along the neuromere boundary region of the diencephalon is regulated by semaphorin 3F and netrin-1. Mol Cell Neurosci 2000; 16(3):206-20.

CHAPTER 8

Semaphorin Signals in Cell Adhesion and Cell Migration:
Functional Role and Molecular Mechanisms

Andrea Casazza,[†] Pietro Fazzari[†] and Luca Tamagnone[†]*

Abstract

Cell migration is pivotal in embryo development and in the adult. During development a wide range of progenitor cells travel over long distances before undergoing terminal differentiation. Moreover, the morphogenesis of epithelial tissues and of the cardiovascular system involves remodelling compact cell layers and sprouting of new tubular branches. In the adult, cell migration is essential for leucocytes involved in immune response. Furthermore, invasive and metastatic cancer cells have the distinctive ability to overcome normal tissue boundaries, travel in and out of blood vessels, and settle down in heterologous tissues. Cell migration normally follows strict guidance cues, either attractive, or inhibitory and repulsive. Semaphorins are a wide family of signals guiding cell migration during development and in the adult. Recent findings have established that semaphorin receptors, the plexins, govern cell migration by regulating integrin-based cell substrate adhesion and actin cytoskeleton dynamics, via specific monomeric GTPases. Plexins furthermore recruit tyrosine kinases in receptor complexes, which allows switching between multiple signaling pathways and functional outcomes. In this article, we will review the functional role of semaphorins in cell migration and the implicated molecular mechanisms controlling cell adhesion.

Introduction

Semaphorins were initially identified as evolutionarily conserved axonal guidance cues in the assembly of the neural circuitry. However, it is now clear that they form a collection of more than twenty soluble or membrane-bound molecular signals (in vertebrates), involved in a range of different processes that often implicate the regulation of cell-substrate adhesion and directional cell migration.

The receptors for semaphorins belong to a family of plasma membrane molecules, the plexins.[1] Two plexins are found in invertebrate genomes, whereas nine genes are known in humans. Interestingly, a common structural domain, known as the "sema domain", is present in the extracellular region of semaphorins, plexins and scatter factor receptors, which underlines the phylogenetic link between these gene families.[2,3] It was recently shown that the structure of the sema domain is a so-called "beta-propeller", similar to that found in α-integrins.[4-6] In addition to the sema domain, the extracellular region of plexins, semaphorins and β-integrins

---

[†]Authors contributed equally to this chapter.
*Corresponding Author: Luca Tamagnone—University of Turin Medical School, Institute for Cancer Research and Treatment (IRCC), Str. Prov. 142, I-10060 Candiolo (Torino), Italy. Email: luca.tamagnone@ircc.it

Semaphorins: Receptor and Intracellular Signaling Mechanisms, edited by R. Jeroen Pasterkamp.
©2007 Landes Bioscience and Springer Science+Business Media.

Table 1. Identified semaphorin receptors

	Semaphorin	Receptor(s)
Secreted	Sema3A	Neuropilin-1 + Plexins (PlexinAs, PlexinD1)
	Sema3B	Neuropilin-1/Neuropilin-2 + Plexins?
	Sema3C	Neuropilin 1/Neuropilin-2 + Plexins (PlexinA??)
	Sema3E	PlexinD1, Neuropilins?
	Sema3F	Neuropilin-2 + Plexins (PlexinA3)
Membranebound	Sema4A	TIM2 (low affinity)
	Sema4D	PlexinB1 (PlexinB2), CD72 (low affinity)
	Sema5A	PlexinB3, Proteoglycans (via TSP repeats)
	Sema6A	PlexinA4
	Sema6B	PlexinA4?
	Sema6D	PlexinA1
	Sema7A	PlexinC1, Integrin-beta1 (affinity unknown)
Viral (secreted)	A39R	PlexinC1

contains cysteine-rich "PSI" domains, further underlining the structural similarity between semaphorin receptors and integrin receptor complexes. The cytoplasmic domain of plexins, instead, lacks any striking homology to known proteins or functional motifs, but it contains two amino acid stretches weakly similar to GTPase Activating Proteins (GAPs), and it was recently shown to catalyse the inactivation of R-Ras monomeric GTPase.[7] In addition to that, plexins are found in association with receptor- and nonreceptor-type tyrosine kinases.[8]

Many secreted class 3 semaphorins are unable to interact with plexins directly. Therefore plexins form receptor complexes with neuropilins, which provide a high-affinity binding site for secreted semaphorins.[1,9] Notably, neuropilins also interact with vascular endothelial growth factors (VEGFs) and with their tyrosine kinase receptors (VEGF-Rs). Although neuropilins carry a small conserved cytoplasmic tail, there are contradictory results on the signaling competence of this domain. According to many reports, neuropilins only provide a ligand-binding platform in the complex and the intracellular signaling is mediated by the associated plexins; other findings, however, suggest an independent signaling function of the intracellular domain of neuropilins.[10] Intriguingly, certain semaphorins have been reported to interact with other receptors, in addition to plexins and neuropilins. An up-to-date index of the identified semaphorin receptors can be found in Table 1. Moreover, transmembrane semaphorins (subclasses 1, 4, 5 and 6) may signal in a bi-directional manner, since they carry intracellular domains that can mediate so-called "reverse" signaling into expressing cells (in addition to the classical "forward" signaling elicited in cells expressing the receptors).[11,12]

Functional Role of Semaphorins in Cell Adhesion and Cell Migration

Initially, most studies on semaphorins focused on their repelling activity for extending axons. Subsequently, it was demonstrated that they regulate the migration of a variety of cells, including neuronal precursors, neural crest cells, oligodendrocytes, endothelial cells, leucocytes, epithelial cells and tumor cells derived from carcinomas, melanomas, etc.[8] As summarized in Table 2, several semaphorins can negatively regulate integrin function, cell adhesion and cell migration. However, semaphorin signaling is multifaceted, and a subset of these ligands (e.g., Sema4D, Sema6D and Sema7A) was furthermore shown to elicit integrin activation/cell-substrate adhesion, axon outgrowth and cell chemotaxis in distinctive conditions (see Table 2). Although the molecular mechanisms responsible for these antagonistic activities have not been completely understood, they seem to implicate distinctive signaling pathways, depending on the targeted cells and on the different components of semaphorin receptor complexes.[8,13]

Table 2. Semaphorin activities in the control of cell migration and cell adhesion

Cell Type	Semaphorin	Receptors (Implicated)	Activity (Observed)	Refs.
Neural crest cells	3A, 3F	NP1, NP2	Repulsion in migration (stripe assay in vitro); repulsion, altered positioning (localized expression in vivo)	28,29,31
DRG growth cones	3A	NP1?	Loss of substrate adhesion	122
DRG neurons	3A	PlexinAs + NP1?	Inhibition of integrin-mediated adhesion and spreading; inhibition of integrin-mediated axonal outgrowth	110
GABAergic neurons	3A, 3F	NP1/NP2?	inhibition of migration, repulsion	26
Platelets	3A	NP1?	Inhibition of integrin-mediated adhesion and spreading; disassembly of F-actin	123
Monocytes	3A	Not neuropilin-1?	Inhibition of migration	124
PAEC endothelial cells	3A	NP1	Inhibition of motility and capillary sprouting (aortic ring assay in vitro)	40
HUVEC endothelial cells	3A (autocrine) 3F	NP1-PlexinA1	Inhibition of integrin activation and substrate adhesion; inhibition of directional migration	16
MDA breast cancer cells	3A autocrine	NP1-PlexinA1	Inhibition of migration (VEGF antagonist)	125
HaCaT keratinocytes	3A	NP1	Inhibition of migration	121
MCF-7 and C100 mammary cancer cells	3F	NP2?	Inhibition of cell adhesion and cell spreading; inhibition E-cadherin-mediated cell adhesion	126,127
NSCLC lung cancer	3F	NP2?	Integrin inhibition	71
SM melanoma cells	3F	Neuropilin 2	Reduced integrin expression, reduced adhesion and migration	43
PAEC endothelial cells	3F	Neuropilin 2	Chemorepulsion	43
HUVEC endothelial cells	3F	NP2?	Inhibition VEGF- and FGF-induced proliferation and angiogenesis (in vivo matrigel plug)	42
Intersomitic vessels	3E	PlexinD1	Inhibition of vascularization (localized in vivo expression)	41
SVEC4-10 endothelial cells	3E	?	Increased migration	67

continued on next page

Table 2. Continued

Cell Type	Semaphorin	Receptors (Implicated)	Activity (Observed)	Refs.
PC12 neuroblasts	3E	?	Neurite outgrowth	67
SKBR3 mammary carcinoma cells	4D (soluble)	PlexinB1	Inhibition of cell adhesion, cell spreading and cell migration; disassembly of adhesive complexes, disassembly of F-actin cables	68
Monocytes, dendritic cells	4D (soluble)	PlexinB1/PlexinC1	Inhibition of migration	124,58
MLP29 liver progenitor cells	4D (soluble)	PlexinB1 + Met	Induction of cell migration, cell survival, branching morphogenesis, invasive growth	70
PAEC or HUVEC endothelial cells	4D (soluble)	PlexinB1 (+Met?)	Induction chemotaxis and angiogenesis (in vitro and in vivo assays)	45,46
Endothelial cells	6A (soluble ectodomain)	Plexins?	Inhibition of migration (VEGF antagonist)	44
Cerebellar granule cells	6A (transmembrane)	Plexins?	Cell-contact repulsion, dissociation of neuronal clusters, initiation of migration	27
Endocardial cells (conotruncal segment)	6D	PlexinA1 + OTK	Attract migration	48
Endocardial cells (ventricular segment)	6D	PlexinA1 + VEGFR2	Repel migration	48
Myocardial cells	6D	PlexinA1 (forward and reverse signaling)	Induced migration and invasive growth	12
Olfactory bulb neurons	7A (soluble)	Integrin β-1	Integrin activation and integrin-mediated axon outgrowth	21
MC3T3 osteoblastic cells	7A (soluble)	Plexins? Integrin β-1?	Increased migration	128
Monocytes	7A (soluble)	?	Increased migration	57
Dendritic cells	A39R	PlexinC1	Inhibition of integrin-mediated adhesion and spreading; inhibition of chemokine-induced migration	59

A list of reports where different semaphorins were shown to regulate the migration or integrin-mediated adhesion of cells expressing endogenous receptors, either by in vitro assays or by localized expression in vivo. Reports concerning other functions of semaphorins are not included.

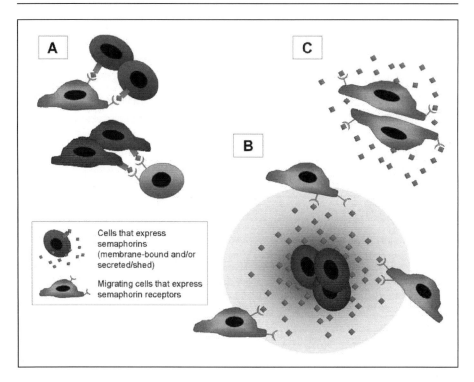

Figure 1. Semaphorin signals act in juxtacrine, paracrine and autocrine manner. A) Membrane-bound semaphorins expressed by guidepost cells can mediate localized signals to guide migration (e.g., Sema6D, Sema6A); B) secreted/shed molecules can diffuse in the area of expressing cells, and demarcate it as a "no entry" zone for cell populations expressing the specific receptors (e.g., Sema3A, Sema3F); C) migrating cells may establish autocrine loops of secreted/shed semaphorins, to police their own migration (e.g., Sema3A).

As mentioned above, semaphorins include secreted and membrane-bound molecules. In addition, certain membrane-bound semaphorins (e.g., Sema4D) are proteolytically released ("shed") in the extracellular space in active form.[14,15] Available data indicate that the control of cell migration by semaphorin signals follows three paradigms, illustrated in Figure 1: (A) membrane-bound semaphorins at the cell surface can mediate localized signals to guide migration; (B) secreted/shed molecules can diffuse in the area of expressing cells, and demarcate it as a "no entry" zone (or instead form a chemotactic gradient) for cell populations expressing the specific receptors; (C) migrating cells may establish autocrine loops of secreted/shed semaphorins, to police their own migration: e.g., it was found that the inactivation of autocrine Sema3A signaling in endothelial cells results in increased directional persistence.[16]

What are the molecular pathways underlying the control of cell adhesion and cell migration by semaphorins? It is known that integrin-mediated adhesion and cell migration are intimately linked processes, since the latter crucially depends on a dynamic regulation of cell tethers to the extracellular matrix (ECM); for a more general review on this topic, see reference 17. Thus, migration must be considered a cyclic multistep process. Migrating cells continuously sense extracellular guidance cues at their polarized "leading edge". These signals, including the semaphorins, may either support further movement in a specific direction ("attractive" cues) or prevent it ("repelling" cues). In response to a repelling cue, the leading edge collapses and small protrusions test the environment, until the cell gets polarized towards a new "permissive" direction. Therefore, repelling cues not only define "off" territories where the migration of a specific cell is not allowed, but also serve to put directional migration "on hold", and reset the threshold for attractive signals. In contrast, when stable adhesive complexes are established, the cell body is pulled forward by cytoskeletal contraction; this results in cell translocation and restarts the extension cycle. It should be underlined that cell migration requires a dynamic regulation of cell-substrate adhesion; in fact, it is hampered both in the presence of stiff adhesive structures, and when integrin function is impaired and adhesive structures cannot form or cannot connect with the actin cytoskeleton. For a long time, semaphorins have been shown to elicit depolymerization and remodeling of the actin cytoskeleton, thereby blocking leading edge extension and inducing lamellipodia retraction (often indicated as the "collapsing response"). Accumulating evidence now shows that semaphorins can negatively regulate integrin function and cell-substrate adhesion. By this means, they can greatly affect cell and axon navigation, since the leading edge of a migrating cell (or an axon growth cone) is steered not only by "polarizing" attracting signals and "depolarizing" repelling signals, but also by changes in the "permissive" cues locally provided by the adhesive substrate.

We will now discuss the experimental data supporting the role of semaphorins in the regulation of cell adhesion and cell migration in vivo, whereas in the second part of the review we will sketch a picture of the signaling pathways currently implicated in these functions.

Axon Guidance

Semaphorins have a prominent role to wire the neural network by guiding axon pathfinding. Notably, the mechanisms controlling axonal extension share remarkable similarities with those acting in cell migration. In fact, at the tip of the axon, a specialized structure termed the growth cone explores the surrounding environment, via repeated cycles of protrusion and collapse. The mechanistic explanation of this process classically focused on the regulation of cytoskeletal dynamics. However, it is well known that integrin-mediated adhesion does play a role in axon guidance, and guidance cues may regulate the affinity of axons to ECM components;[18,19] for a review see reference 20. Nonetheless, direct links between semaphorins and integrins during axon guidance were provided only recently. By a series of elegant experiments in vitro and in vivo, Pasterkamp and colleagues showed that the GPI-linked Sema7A is necessary to promote axon outgrowth from the olfactory bulb.[21] In this model, Sema7A elicits the activation of integrin-β1 and promotes axon outgrowth via MAPK activation downstream to integrin signaling. In principle, semaphorins known to inhibit integrin activation, such as Sema3A and Sema3F, may act in an opposite manner: thereby preventing integrin activation and growth cone adhesion in areas from where the axon is steered away. Direct evidence of this mechanism, however, is currently missing.

An intriguing insight into the connection between semaphorins and integrin-mediated adhesion in axon guidance emerges from the functional role of Sema3B in the formation of the anterior commisure (AC). This is a bipartite tract formed by an anterior (ACa) and a posterior (ACp) limb that converge into a common fascicle during development. Sema3B exerts a dual function on these axons: being attractive for ACa and repulsive for ACp.[22] Although the implicated mechanisms are not fully understood, it was found that FAK, a Src-like tyrosine kinase associated with integrin-based focal adhesions, is recruited to the membrane and gets phosphorylated only in ACa neurons (those attracted by the semaphorin). Furthermore, pharmacological inhibition of Src, likely implicated in FAK activation, converts the attractive response of ACa neurons into repulsion.[22]

During development, axonal growth cones are eventually guided by specific signals to connect with target cells and form synapses, later to be remodeled in the adult. Pre and postsynaptic membranes are actually linked by cell junctions, similar to those found between epithelial cells and between immune-response cells (the "immunological synapse"). A role for semaphorins in the formation and plasticity of synapses was initially proposed in Drosophila, where both secreted Sema2 and the transmembrane Sema1a were found to regulate synaptogenesis, through different mechanisms.[11,23] In mammals, during hippocampal development, exuberant synaptic connections are normally eliminated, followed by the retraction of redundant "nonconnected" axons. Noteworthy, in mice lacking PlexinA3 or NP2, the pruning of unnecessary synapses and axon collaterals is impaired.[24] Therefore, semaphorins not only act as topographical guiding cues, but can also provide temporally-restricted signals controlling the stability of cell-cell contacts, for instance in synaptic plasticity. Intriguingly, since the expression of many semaphorins is maintained in the adult nervous system, these findings could pave the way for studying the potential role of semaphorin signaling in synaptic plasticity disorders.

Neuronal Migration

During the development of the central nervous system (CNS), cell migration is a fundamental process, since neurons and glial cell usually originate in proliferative zones that are far from the sites where they will ultimately reside. The role of the semaphorins in the control of this migratory process is currently emerging.

Most of the neurons migrate radially, using "radial glia" as a track, and the regulation of this process is believed to be essentially contact-mediated. A significant population of GABAergic neurons, instead, arises in the ventricular zone of the subpallium and undergoes tangential migration. While a small fraction of these cells is then sorted to the striatum, the majority avoids this region, ending their journey in the piriform cortex and the neocortex. In vivo and ex vivo experiments[25,26] showed that the striatum expresses Sema3A and Sema3F. The neurons avoiding this area express both neuropilin-1 and -2, while those invading the striatum do not express any competent semaphorin receptors. In fact, loss of neuropilin function hijacks the control of migration into the striatum, thus increasing the number of neurons entering this area and reducing the number of those migrating to the cortex.[25] In sum, neuropilin expression segregates different subpopulations of GABAergic neurons, and semaphorin signaling is required for their sorting toward specific locations.

Another example of semaphorins controlling neuronal migration is found in cerebellar development. Here granule cells proliferate after birth in the external granular layer (EGL), from where they start an inward radial migration, through the molecular layer (ML) and on the "track" of Bergmann glia, into the internal granule layer (IGL). The role of the transmembrane semaphorin Sema6A in this process was recently demonstrated by gene-ablation experiments, providing a fascinating example of the fine-tuned clockwork of neural development.[27] In Sema6A-deficient mice, the inward migration of granule cells is impaired as many of them remain ectopic in the ML (where they differentiate and connect with the mossy fibers). An extensive analysis of these mutants revealed that, surprisingly, granule cell development is otherwise completely normal (e.g., concerning proliferation, differentiation, survival and even neurite extension). The authors suggest a model in which Sema6A acts as cell-cell contact repellent, providing the cell body of granule cells with a "go" signal to leave the EGL and start the inward radial migration. Notably Sema6A expression in granule cells is turned off as soon as they leave the EGL.

Neural Crest Cell Migration

Neural crest cells originate in the dorsal region of the neural tube and migrate out to peripheral districts. They are in fact the precursors of a remarkable assortment of differentiated cell types, including pigment cells (melanocytes), skeletal cartilage cells (chondroblasts), cells "patterning" heart development, autonomic ganglion cells, and glial cells of the peripheral nervous system. The analysis of these cells in vitro provided one of the first examples that semaphorins can regulate cell migration, beyond axon guidance.[28] Studies in chicken embryo have later revealed that neural crest cells in the hindbrain and in the trunk express neuropilin receptors, and are selectively excluded from those regions containing Sema3A and Sema3F.[29] Moreover, ectopic expression of Sema3A or Sema3F in vivo (or expression of soluble neuropilins as decoy receptors) results in the disruption of the normal "streams" of migrating cells. These findings are consistent with loss-of-function experiments in mice. In fact, both Sema3A and NP1 knock-out animals display defects in the migration of neural crest cells of the sympathetic lineage, leading to the mislocalization of mature sympathetic neurons and to an impaired development of sympathetic ganglia.[30] Similar functions were recently reported for Sema3F in mouse and in zebrafish development.[31,32] In addition, in experiments on explanted sympathetic ganglia, Sema3A, possibly due to its repelling activity, promoted cell aggregation and tight neurite fasciculation.[30] Notably, a subset of migrating neural crest cells characterized by the expression of PlexinA2 is crucially implicated in cardiac development[33,34] (see below).

Oligodendrocyte Migration

As mentioned above, cell guidance is fundamental for migrating glial precursors too. Although little is known about the specific cues controlling this process, it is assumed that some of them may be common to axon pathfinding and neuronal cell migration. In fact, there is evidence indicating a potential role of different semaphorins in oligodendrocyte migration.[35] Moreover, it was reported that oligodendrocyte precursors recruited to the optic nerve express both neuropilin-1 and -2; interestingly, a subpopulation of these cells is repelled by Sema3A, whereas Sema3F acts as chemoattractant and mitogenic factor.[36] Other reports indicated a specific expression of transmembrane semaphorins Sema5A and Sema4D in oligodendrocytes, suggesting a role in the regulation of axon pathfinding, myelination and nerve regeneration.[37,38]

Angiogenesis

There is consistent evidence that semaphorin signals regulate endothelial cell migration and angiogenesis (for details, see an associated review by Neufeld and coworkers in ref. 39). Sema3A was shown to inhibit both the adhesion of endothelial cells to the extracellular matrix and their directional migration.[16,40] Moreover, Sema3A- and Sema3E-deficient mice display defects in vascular development,[16,41] and Sema3F and Sema6A were shown to act as anti-angiogenic factors in experimental tumors.[42-44] On the other hand, Sema4D (and other members of subclass 4) were consistently reported to have pro-angiogenic activity in vitro and in vivo.[45,46]

Organ Morphogenesis

The role of Sema3A, Sema3C and Sema6D in cardiac development is well documented in vivo by gene-loss experiments. Secreted semaphorins Sema3A and Sema3C regulate the migration of neural crest cells that play a crucial role in shaping the heart and the aortic outflow in mice.[33,34,47] In chick, Sema6D signaling was found to affect the organogenesis of both the ventricle and the cardiac outflow tract. Interestingly, the transmembrane semaphorin Sema6D acts bidirectionally: by forward signaling via PlexinA1, and by "reverse" signaling, via the association of Abl-regulatory proteins to its cytoplasmic tail.[12,13,48] Moreover, forward signals mediated by Sema6D-PlexinA1 interaction can either be attractive or repelling in different cell populations, depending on the receptor tyrosine kinase associated with the plexin, i.e., VEGFR2 or OTK, respectively. On the other hand, "reverse" signals elicited into Sema6D-expressing cells seem to promote invasive growth. For more details on this topic, see reference 49.

The morphogenetic role of semaphorins has also been extensively addressed in lung development. Moreover, lung development may provide a general example of tubulogenesis and branching morphogenesis, as observed in other organs (e.g., kidney, pancreas etc.). This process begins from a simple epithelial sheet that ends up as a branched tubular structure. Notably, the remodelling of epithelial layers somewhat parallels the migration process of a single cell, since the entire cluster trails the movements of the cells at the leading edge, sensing the environmental guidance cues. In analogy to neural development, we speculate that semaphorins act by defining permissive and restrictive regions during branching morphogenesis, to carve the shape of the epithelial tree. In fact, at the early stage of lung organogenesis (embryonic days 11-13) Sema3A is expressed in the distal mesenchyme, while neuropilins are found in the epithelium of terminal buds. It was shown in organotypic cultures that Sema3A inhibits the branching of lung epithelium, while Sema3C and Sema3F increase the branching and slightly promote epithelial cell proliferation.[50,51] These data support a model in which the differential distribution of semaphorins, in concert with that of growth factors and other morphogenetic cues, coordinates the shaping of the lung epithelial tree.

Numerous reports have indicated a developmental role of semaphorin signals in preventing ectopic cell contacts during epidermal morphogenesis in the nematode *C.elegans*; in particular, semaphorins are implicated in the sorting of cells into distinct rays of the developing male tail.[52-54] Moreover, although the interpretation of the morphogenetic process differs, two recent reports have established a role of Sema1a/PlexinA signaling in regulating the developmental migration of vulval precursor cells in the same organism.[55,56]

Leucocyte Migration

Scattered reports have shown that semaphorins can affect leucocyte migration, as assayed by in vitro migration assays.[57-59] However, functional in vivo evidence of this activity is still lacking. On the other hand, knock-out mouse models generated in H. Kikutani's lab have clearly demonstrated that class 4 semaphorins Sema4D and Sema4A regulate the immune response mediated by lymphocytes and dendritic cells.[60,61] Interestingly, PlexinA1 was shown to have a role in the "adhesion" between lymphocytes and regulatory cells[62] (the so called "immunological synapse"), inviting the idea that semaphorins could police this association. This matter is reviewed in detail in an associated review.[63]

Tumor Cell Migration

Tumor cell migration is a fundamental process in cancer progression, as it sustains local invasion and metastatic dissemination. Notably, it depends not only on the activity of motogenic factors and guidance cues, but also on mechanisms dissolving the extracellular matrix and inhibiting tumor cell apoptosis. Several scattered reports have indicated an altered expression of semaphorins in cancer (inter alia see refs. 64,65). Experimental evidence in vitro and in vivo supports these data, suggesting that semaphorin signaling can act to promote or to inhibit cancer progression, depending on the receptor complex involved.[42-44,66,67] One interesting example is Sema4D, which was shown to inhibit the migration of carcinoma cells expressing the receptor PlexinB1,[68] while it sustains the constitutive activation of the oncogenic receptor Met (associated with the plexin) in other cells.[69,70] Noteworthy, semaphorins found to inhibit the migration of tumor cells, also affect integrin-dependent adhesion or integrin expression.[43,68,71] The effect of tumor-produced semaphorins in vivo is further explained by an interference with cancer-associated angiogenesis.[42-44] For a detailed review of this exciting field of ongoing research, see reference 39.

Signaling Molecules Mediating Semaphorin Function in Cell Adhesion and Migration

The cytoplasmic domain of plexins, or SP domain, is split into two highly conserved regions, separated by a "linker domain" with more divergent sequence.[72] This domain lacks any striking homology to known proteins or functional motifs; however it includes short sequences with similarity to GTPase activating proteins[73] (or GAPs), and it was recently shown to mediate GAP activity for R-Ras, leading to the inactivation of this small GTPase.[7] Figure 2 shows many intracellular signal transducers that have been implicated in plexin-mediated functions, some of which interact directly with the *SP* cytoplasmic domain.[74] However, it remains largely unclear how these different pathways connect to mediate functional responses to semaphorins. In this review, we will focus on those molecules known to have a role in the control of cell adhesion and cell migration.

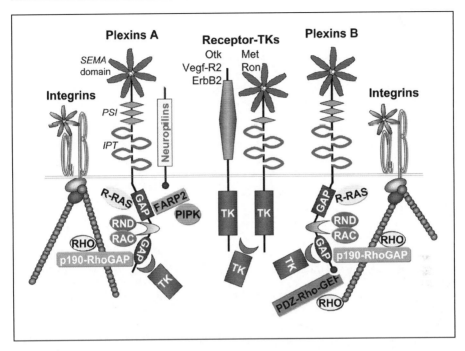

Figure 2. Plexins associate with a range of signal transducers implicated in the control of integrin function and cytoskeletal dynamics. The figure schematically depicts most of the players implicated in the control of cell adhesion and cell migration mediated by semaphorins (see the text for details). Plexins and Receptor Tyrosine Kinases of the Met family share a common structural domain known as the "sema domain" (shown in red), whose structure is a "beta-propeller", similar to that found in α-integrins. In addition to the sema domain, the extracellular region of plexins contains two/three PSI domains (found in Plexins, Semaphorins and β-Integrins; shown in green) and three/four IPT domains (Ig-like, Plexins, Transcription factors; shown in blue), which are putative protein-protein interaction motifs. The cytoplasmic domain of plexins contains two conserved regions with intrinsic R-Ras-GAP activity (depicted in blue), and a linker segment more variable in the family, which interacts with small GTPases, such as Rnd1 and Rac1 (shown in yellow). The cytoplasmic domain of plexins furthermore interacts with p190-RhoGAP, controlling the actin cytoskeleton, and with FARP2, which sequesters and inactivate PIPKIgamma (resulting in the destabilization of cell adhesive complexes). At the C-terminus of the plexins of B subfamily there is a consensus sequence for PDZ domains, which mediates the selective association with PDZ-RhoGEFs (see the text for details). A color version of this figure is available online at http://www.Eurekah.com.

Monomeric G Proteins As Main Regulatory Switches for Plexin Signaling

Monomeric GTPases of the Rho and Ras family are pivotal regulators of intracellular cell signaling. Through their activity, specific signals from the extracellular microenvironment may be rapidly translated into cytoskeletal remodeling events, such as those regulating cell adhesion and cell migration. It is therefore not surprising that these molecules play a pivotal role in semaphorin signaling (for more details see ref. 75).

It was demonstrated that the cytoplasmic domain of plexins directly associates with GTPases of the Rho family and with their regulatory proteins. For instance, activated GTP-bound Rac1 was found to interact directly with the cytoplasmic domain of PlexinB1 and PlexinA1.[76-78]

However, plexins were not found to mediate GAP activity for Rac. Moreover, previous reports showed that both the expression of dominant negative Rac[79,80] and the treatment with an inhibitory peptide for Rac[81] block Sema3A-induced neuron growth cone collapse, indicating the requirement for Rac activation in plexin signaling. Therefore, in an effort to explain its functional role, this interaction was proposed to: (1) direct the localization of active Rac to the cell membrane, where it could regulate cytoskeletal remodeling and protein trafficking;[82] (2) locally sequester the active form of the GTPase from its downstream targets, such as PAK1, thereby inhibiting Rac signaling at the cell membrane;[83,84] (3) mediate a conformational change in the cytoplasmic domain of the plexin, regulating its signaling activity or further interactions with other cytoplasmic proteins;[78] (4) control the targeting of plexin molecules at the cell surface.[84,85]

Independent reports showed that another GTPase of the Rho family, Rnd-1, which is constantly loaded with GTP due to low intrinsic GAP activity, can associate with the same region of the intracellular domain of Plexins as Rac.[86,87] However, similar to Rac, plexins do not carry any GAP activity to inactivate Rnd1. Instead, the association of this GTPase to the cytoplasmic domain of plexins seems to act as a switch, to allow for further interaction with other transducers, such as R-Ras or GTP/GDP exchanger proteins (or GEFs) for RhoA.[7,87] Notably, mutations in the primary sequence of this GTPase-binding region of plexins also affect cell surface expression of the receptor, by unknown mechanisms.[85] In conclusion, the interaction with Rac1 and Rnd1 seems to play a major regulatory role upstream to plexin signaling, by policing receptor activation and interaction with intracellular transducer molecules.

Further extracellular and intracellular mechanisms have been reported to provide regulatory switches for plexin signaling. For example Ig-like cell adhesion molecules (such as L1-CAM or Nr-CAM) associate with neuropilins and may differentially regulate certain semaphorin receptor complexes, depending on cell contact-mediated homophilic interactions with neighbouring cells.[88] Other intriguing findings implicate that the differential expression of proteoglycans in diverse areas of the nervous system may assign either attractive or repelling function to Sema5A signals.[89] Furthermore, at the intracellular level, it was shown that the balance between the cyclic nucleotides cAMP and cGMP may switch Sema3A signaling from repulsion to attraction.[90]

Integrins and the Actin Cytoskeleton Are Major Targets of Plexin Signaling

Several reports show that plexin signaling primarily regulates integrin function. In most instances, semaphorin stimulation leads to: decreased cell-substrate adhesion, decreased integrin expression, block of integrin activation or transition of integrins in the inactive state, disassembly of integrin-based cell adhesion complexes, actin cytoskeleton remodeling with loss of stress fibers, and morphological rearrangements such as retraction of cell protrusions and cell rounding (see Table 2). In addition, loss of integrin activation (and of the associated "outside-in" signaling) may explain other functional effects induced by semaphorins, such as reduced proliferation rate and cellular apoptosis. In certain instances, however, semaphorins were reported to regulate integrin function in the opposite manner: for example, Sema7A stimulation induces the activation of β-integrins and of their associated intracellular pathway, including FAK and MAPK.[21]

The molecular mechanisms of integrin regulation by semaphorins are partly understood. At present, a compelling evidence that semaphorins or their receptors can directly interact with integrins is missing. However, plexins carry an intrinsic GAP activity for active R-Ras, a monomeric GTPase known to induce integrin activation.[7] Notably, two conserved arginine residues are essential for R-Ras inactivation by plexins, and their point mutation results in the loss of most (although not all) functional responses mediated by plexins.[68,73,91] R-Ras is a major positive regulator of integrin function:[92] it gets activated upon cell contact to the ECM,[93] and in turn it sustains the clustering and activation of β1-integrin receptors (as well as others), it enhances focal adhesion formation and cell adhesion, and it regulates positively cell spreading, cell migration and invasion, as well as axon outgrowth (for a general review see ref. 94). These functions qualify R-Ras as the ideal target for the inhibitory activity of semaphorins; nonetheless the molecular mechanisms involved in this pathway are presently unclear. Phosphatidylinositol-3 kinase (PI3K) and its main product PIP3 have been initially implicated as major effectors of R-Ras,[95,96] to elicit the recruitment at the cell membrane of exchange factors activating Rac1, thereby promoting integrin-mediated adhesion. However, two recent reports demonstrated that R-Ras signaling instead induces RhoA activation and Rac1 inactivation at the leading edge, via still unknown pathways.[93,97] Therefore, further studies are required to elucidate the mechanism whereby R-Ras inactivation mediated by semaphorins may regulate cell adhesion and cell migration.

Interestingly, other findings indicate that RhoA regulation is fundamental in plexin signaling. For instance, p190RhoGAP, a major negative regulator of RhoA,[98] was recently shown to associate with plexins and become activated upon semaphorin binding.[99] Moreover, in p190-deficient cells, plexin signaling was abrogated or significantly hindered.[99] The regulatory function of RhoA in cytoskeletal remodelling, cell adhesion and cell migration is rather complex and multifaceted.[100,101] During cell migration, the levels of active RhoA are highly dynamic: they are cyclically downregulated at the leading edge to disassemble the actin cytoskeleton and release cell-substrate adhesions, and this is transiently required to form cell protrusions. However, once lamellipodia establish new integrin-mediated contacts, RhoA becomes activated and elicits the formation of thick actin cables connected to the adhesive complexes, whose contraction is the "pulling" force in cell body translocation. Therefore, RhoA inactivation mediated by plexins inhibits acto-myosin contractility, disassembles actin cables and leads to weak and unstable adhesive complexes, which in turn blocks directional migration.[68] Moreover, activated p190 is known to recruit to the complex p120-RasGAP, a general downregulator of Ras-like GTPases, including R-Ras;[94] this could provide an additional mechanism to mediate R-Ras inactivation upon plexin signaling, and perhaps a functional link between the two pathways. Noteworthy, since data suggest that semaphorin signaling requires the regulation of both R-Ras and RhoA, further investigation is needed to establish the specific role of these GTPases in different functional responses in vivo. For instance, recent findings indicate that R-Ras expression in vivo is mostly confined to blood vessels,[102] which could underline a predominant role of this pathway in semaphorin-mediated control of angiogenesis.

It was also shown by many groups that plexins of the B subfamily can associate with a subfamily of RhoA GTP/GDP exchangers (PDZ-RhoGEFs), via a unique and conserved C-terminal sequence that is not found in other plexins.[103-107] This interaction is not influenced by the ligand, but depends on the association of Rnd1 to the plexin[87] and on the activity of ErbB2 tyrosine kinase recruited in the complex.[108] Therefore, in certain instances, the activation of plexins B can induce RhoA activation, which is likely to be involved in specific functional responses mediated by plexins of the B subfamily, such as the pro-angiogenic activity of Semaphorin 4D.[45] Plexin-associated RhoGEFs may furthermore be important in axon retraction, as demonstrated by expressing dominant negative forms of the RhoGEF or inhibiting Rho-dependent kinase (ROCK) in neurons.[106] Moreover, genetic evidence in Drosophila showed that plexin signaling in axon guidance implicates Rho activation.[83] Intriguingly, PDZ-RhoGEFs associated with plexin Bs are multi-domain proteins that could provide a selective platform for the interaction with additional independent signaling pathways.[109]

Recent data highlighted the role of FARP2 (a FERM domain protein) to mediate Sema3A-mediated integrin inhibition in neurons derived from dorsal root sensory ganglia (DRG).[110] Moreover, in FARP2-depleted neurons, Sema3A-mediated repulsion of neurite outgrowth is suppressed. Noteworthy, FARP2 was found to interact with plexins of the A subfamily, but not with PlexinB1. This molecule is likely to act by suppressing the kinase activity of Phosphatidylinositol-4-Phosphate 5-Kinase Type Iγ(PIPK-$γ_{661}$), which is associated with integrin-based adhesive complexes. Intriguingly, a previous report indicated that Sema3A-mediated neurite retraction requires instead the activity of another isoform of PIPK, called PIPK-Iα.[111] Although not localized to focal contacts, PIPK-Iα activity was shown to induce vinculin dissociation from the adhesive complexes and release substrate adhesion.[111] Therefore, the regulation of PIP2 levels at the cell surface seems to be an important mechanism to account for semaphorin-mediated control of cell substrate adhesion and actin remodelling.

Plexins and Tyrosine Kinases: Receptor Complexes That Can Make the Difference

Tyrosine phosphorylation is another major regulatory mechanism implicated in cell adhesion and cell migration. Plexins do not carry an intrinsic kinase activity, however they can trigger the activation of plexin-associated receptor-type and nonreceptor-type tyrosine kinases (for a recent review, also see ref. 112). For example, PlexinB1 was found in complex with the receptor-tyrosine kinases Met, Ron and ErbB2.[69,70,108] Upon semaphorin binding, the latter become activated, and in turn they phosphorylate plexins on tyrosine residues. The functional role of these interactions is currently under investigation: they could play a regulatory role on the cytoplasmic domain of plexins, but also trigger independent motogenic and survival pathways such as those mediated by Rac1 and RhoA, PI3K-AKT, FAK/MAPK, etc. This may also explain the bifunctional activity of Sema4D, which can either inhibit or promote cell adhesion and cell migration, depending on the activation of the associated Met kinase.[69,113]

Another interesting example is PlexinA1, which was found in complex with either VEGFR2 or the catalytic-inactive tyrosine kinase receptor Otk, in different cell populations.[48] Upon binding Sema6D, this plexin elicts alternative signaling pathways in different cells: those expressing Otk were repelled by the semaphorin, while those expressing VEGFR2 were attracted. Notably, Otk is an orphan receptor kinase whose functional role in semaphorin signaling seems to be conserved from invertebrates to humans.[114] In sum, the association with different kinase partners could explain the switch between different functions mediated by semaphorins, by eliciting independent transduction cascades.

Several nonreceptor-type tyrosine kinases have been implicated in semaphorin signaling, many of which can be found in focal adhesive complexes (e.g., Fyn, FAK, PYK2 and Src). In particular, Gutkind and coworkers have demonstrated that—upon Sema4D stimulation—the specific receptor PlexinB1 recruits PYK2 and Src cytosolic tyrosine kinases, eventually leading to the activation of p110-PI3K and of its downstream pathway.[91] Notably, this pathway is independent of the ability to inactivate R-Ras and it is antagonistic to that reported in response to Sema3F.[115] Moreover, there is evidence that it contributes to the chemotactic activity and pro-angiogenic function of Sema4D in endothelial cells.[91]

Some Open Questions

Semaphorins are known to trigger multiple signaling pathways, via plexins or alternative (poorly defined) mechanisms. Moreover, semaphorin stimulation may lead to different functional outcomes, depending on plexin-associated components in the receptor complex. A currently unanswered question is: how is the function of these semaphorin receptor complexes regulated in cells?

Although there is clear evidence that semaphorin signals regulate cell adhesion, it is almost completely unknown whether different integrin complexes are regulated by semaphorin receptors in the same manner. Could semaphorin signals elicit a differential adhesion to various ECM components found in tissues?

The functional link between regulation of integrin function and cell migration deserves further investigation. For instance, we speculate that the effects on cell migration elicited by semaphorins may depend on the grade of inhibition of integrin-mediated adhesion. Thereby, a limited release of cellular tethers to the ECM could actually be synergistic with motogenic signals.

The role of various small GTPases in semaphorin-mediated control of cell migration is less than clear, and some conflicting evidence has been obtained in different experimental models.

Notably, certain "endpoint" functional responses in vitro, such as cell or growth cone collapse, may not reflect the dynamic nature of cell migration and axon guidance in vivo. Moreover, it is known that interfering with the levels of expression or activity of small GTPases may produce unpredictable basal effects. Thus, a more reliable analysis of the function of these molecules in plexin signaling would require the real-time detection of their sub-cellular localization and activation in living cells (e.g., by fluorescence-based techniques).

The membrane-bound or soluble nature of different semaphorins is often ignored in experimental studies, while this may be relevant to the function in vivo in several ways (localized "clustered" cues vs. diffused gradients, reverse signaling, etc.). Thus, new functional assays need to be developed to better address this issue in vitro.

Due to the seminal work of H. Fujisawa and collaborators, semaphorin receptors were originally described as homophilic/heterophilic cell adhesion molecules.[116,117] Later reports confirmed this observation,[62,118,119] suggesting the speculation that -by interfering with this function- semaphorins may regulate not only cell-substrate adhesion, but also cell-to-cell adhesion. Further analyses are thus required to test this hypothesis.

Finally, one intriguing aspect of semaphorin signaling in vivo is that independent cell types within a tissue may express receptors for the same semaphorin, leading to a complex scenario of integrated functional responses. For example, it is now clear that the development of neural and vascular networks share a range of guidance cues (including semaphorins, neurotrophins and VEGFs), which results in the coordinated arborization of vessels and neurites (for a review see ref. 120). Moreover, it is intriguing to note that semaphorins in tumors may exert a double functional role in cancer cells and in tumor-associated stroma. Future studies will tell how signals mediated by semaphorins are functionally integrated in different tissues, during development and in the adult.

Conclusions

Semaphorin signaling is now recognized as a major guidance code for axon pathfinding and cell migration. These activities underlie the functional relevance of semaphorins and their receptors in complex developmental processes, such as the wiring of the neural network, the migration of neural crest cells involved in organogenesis, and the patterning of the vascular network. Furthermore, in the adult, the expression of semaphorins is maintained, and these signals are suspected to have an important role in neural plasticity and regeneration, in the immune function, and in cancer progression. Recent findings have established that the regulation of integrin-mediated adhesion is an important outcome of semaphorin signals, which may well account for their activities in axon guidance and cell migration. The molecular mechanisms linking semaphorins to integrins now begin to be understood; however further studies will be required to dissect the multiple intracellular pathways elicited by these signals.

Acknowledgements

We wish to thank Silvia Giordano, Livio Trusolino and Asha Balakrishnan for critical reading of the manuscript. For the financial support to their research activity, the authors wish to thank the Italian Association for Cancer Research (AIRC), the Regione Piemonte and MIUR-FIRB Neuroscience (Grant RBNE01WY7P).

References

1. Tamagnone L, Artigiani S, Chen H et al. Plexins are a large family of receptors for transmembrane, secreted, and GPI-anchored semaphorins in vertebrates. Cell 1999; 99(1):71-80.
2. Artigiani S, Comoglio PM, Tamagnone L. Plexins, semaphorins, and scatter factor receptors: a common root for cell guidance signals? IUBMB Life 1999; 48(5):477-482.
3. Winberg ML, Noordermeer JN, Tamagnone L et al. Plexin A is a neuronal semaphorin receptor that controls axon guidance. Cell 1998; 95(7):903-916.
4. Antipenko A, Himanen JP, van Leyen K et al. Structure of the semaphorin-3A receptor binding module. Neuron 2003; 39(4):589-598.

5. Love CA, Harlos K, Mavaddat N et al. The ligand-binding face of the semaphorins revealed by the high-resolution crystal structure of SEMA4D. Nat Struct Biol 2003; 10(10):843-848.
6. Gherardi E, Youles ME, Miguel RN et al. Functional map and domain structure of MET, the product of the c-met protooncogene and receptor for hepatocyte growth factor/scatter factor. Proc Natl Acad Sci USA 2003; 100(21):12039-12044.
7. Oinuma I, Ishikawa Y, Katoh H et al. The Semaphorin 4D receptor Plexin-B1 is a GTPase activating protein for R-Ras. Science 2004; 305(5685):862-865.
8. Tamagnone L, Comoglio PM. To move or not to move? Semaphorin signaling in cell migration. EMBO Rep 2004; 5(4):356-361.
9. Takahashi T, Fournier A, Nakamura F et al. Plexin-neuropilin-1 complexes form functional semaphorin-3A receptors. Cell 1999; 99(1):59-69.
10. Wang L, Zeng H, Wang P et al. Neuropilin-1-mediated vascular permeability factor/vascular endothelial growth factor-dependent endothelial cell migration. J Biol Chem 2003; 278(49):48848-48860.
11. Godenschwege TA, Hu H, Shan-Crofts X et al. Bi-directional signaling by Semaphorin 1a during central synapse formation in Drosophila. Nat Neurosci 2002; 5(12):1294-1301.
12. Toyofuku T, Zhang H, Kumanogoh A et al. Guidance of myocardial patterning in cardiac development by Sema6D reverse signaling. Nat Cell Biol 2004; 6(12):1204-1211.
13. Comoglio PM, Tamagnone L, Giordano S. Invasive growth: a two-way street for semaphorin signaling. Nat Cell Biol 2004; 6(12):1155-1157.
14. Elhabazi A, Delaire S, Bensussan A et al. Biological activity of soluble CD100. I. The extracellular region of CD100 is released from the surface of T lymphocytes by regulated proteolysis. J Immunol 2001; 166(7):4341-4347.
15. Wang X, Kumanogoh A, Watanabe C et al. Functional soluble CD100/Sema4D released from activated lymphocytes: possible role in normal and pathologic immune responses. Blood 2001; 97(11):3498-3504.
16. Serini G, Valdembri D, Zanivan S et al. Class 3 semaphorins control vascular morphogenesis by inhibiting integrin function. Nature 2003; 424(6947):391-397.
17. Ridley AJ, Schwartz MA, Burridge K et al. Cell migration: integrating signals from front to back. Science 2003; 302(5651):1704-1709.
18. Ivankovic-Dikic I, Gronroos E, Blaukat A et al. Pyk2 and FAK regulate neurite outgrowth induced by growth factors and integrins. Nat Cell Biol 2000; 2(9):574-581.
19. Turney SG, Bridgman PC. Laminin stimulates and guides axonal outgrowth via growth cone myosin II activity. Nat Neurosci 2005; 8(6):717-719.
20. Nakamoto T, Kain KH, Ginsberg MH. Neurobiology: New connections between integrins and axon guidance. Curr Biol 2004; 14(3):R121-R123.
21. Pasterkamp RJ, Peschon JJ, Spriggs MK et al. Semaphorin 7A promotes axon outgrowth through integrins and MAPKs. Nature 2003; 424(6947):398-405.
22. Julien F, Bechara A, Fiore R et al. Dual functional activity of semaphorin 3B is required for positioning the anterior commissure. Neuron 2005; 48(1):63-75.
23. Matthes DJ, Sink H, Kolodkin AL et al. Semaphorin II can function as a selective inhibitor of specific synaptic arborizations. Cell 1995; 81(4):631-639.
24. Liu XB, Low LK, Jones EG et al. Stereotyped axon pruning via plexin signaling is associated with synaptic complex elimination in the hippocampus. J Neurosci 2005; 25(40):9124-9134.
25. Marin O, Yaron A, Bagri A et al. Sorting of striatal and cortical interneurons regulated by semaphorin-neuropilin interactions. Science 2001; 293(5531):872-875.
26. Tamamaki N, Nakamura K, Kaneko T. Cell migration from the corticostriatal angle to the basal telencephalon in rat embryos. Neuroreport 2001; 12(4):775-780.
27. Kerjan G, Dolan J, Haumaitre C et al. The transmembrane semaphorin Sema6A controls cerebellar granule cell migration. Nat Neurosci 2005; 8(11):1516-1524.
28. Eickholt BJ, Mackenzie SL, Graham A et al. Evidence for collapsin-1 functioning in the control of neural crest migration in both trunk and hindbrain regions. Development 1999; 126(10):2181-2189.
29. Osborne NJ, Begbie J, Chilton JK et al. Semaphorin/neuropilin signaling influences the positioning of migratory neural crest cells within the hindbrain region of the chick. Dev Dyn 2005; 232(4):939-949.
30. Kawasaki T, Bekku Y, Suto F et al. Requirement of neuropilin 1-mediated Sema3A signals in patterning of the sympathetic nervous system. Development 2002; 129(3):671-680.
31. Gammill LS, Gonzalez C, Gu C et al. Guidance of trunk neural crest migration requires neuropilin 2/semaphorin 3F signaling. Development 2006; 133(1):99-106.
32. Yu HH, Moens CB. Semaphorin signaling guides cranial neural crest cell migration in zebrafish. Dev Biol 2005; 280(2):373-385.

33. Brown CB, Feiner L, Lu MM et al. PlexinA2 and semaphorin signaling during cardiac neural crest development. Development 2001; 128(16):3071-3080.
34. Feiner L, Webber AL, Brown CB et al. Targeted disruption of semaphorin 3C leads to persistent truncus arteriosus and aortic arch interruption. Development 2001; 128(16):3061-3070.
35. Cohen RI, Rottkamp DM, Maric D et al. A role for semaphorins and neuropilins in oligodendrocyte guidance. J Neurochem 2003; 85(5):1262-1278.
36. Spassky N, de Castro F, Le Bras B et al. Directional guidance of oligodendroglial migration by class 3 semaphorins and netrin-1. J Neurosci 2002; 22(14):5992-6004.
37. Goldberg JL, Vargas ME, Wang JT et al. An oligodendrocyte lineage-specific semaphorin, Sema5A, inhibits axon growth by retinal ganglion cells. J Neurosci 2004; 24(21):4989-4999.
38. Moreau-Fauvarque C, Kumanogoh A, Camand E et al. The transmembrane semaphorin Sema4D/CD100, an inhibitor of axonal growth, is expressed on oligodendrocytes and upregulated after CNS lesion. J Neurosci 2003; 23(27):9229-9239.
39. Neufeld G, Lange T, Varshavsky A, Kessler O. Semaphorin signaling in vascular and tumor biology. In: Pasterkamp RJ, ed. Semaphorins: Receptor and Intracellular Signaling Mechanisms. Georgetown: Landes Bioscience, 2006.
40. Miao HQ, Soker S, Feiner L et al. Neuropilin-1 mediates collapsin-1/semaphorin III inhibition of endothelial cell motility: functional competition of collapsin-1 and vascular endothelial growth factor-165. J Cell Biol 1999; 146(1):233-242.
41. Gu C, Yoshida Y, Livet J et al. Semaphorin 3E and plexin-D1 control vascular pattern independently of neuropilins. Science 2005; 307(5707):265-268.
42. Kessler O, Shraga-Heled N, Lange T et al. Semaphorin-3F is an inhibitor of tumor angiogenesis. Cancer Res 2004; 64(3):1008-1015.
43. Bielenberg DR, Hida Y, Shimizu A et al. Semaphorin 3F, a chemorepulsant for endothelial cells, induces a poorly vascularized, encapsulated, nonmetastatic tumor phenotype. J Clin Invest 2004; 114(9):1260-1271.
44. Dhanabal M, Wu F, Alvarez E et al. Recombinant semaphorin 6A-1 ectodomain inhibits in vivo growth factor and tumor cell line-induced angiogenesis. Cancer Biol Ther 2005; 4(6):659-668.
45. Basile JR, Barac A, Zhu T et al. Class IV semaphorins promote angiogenesis by stimulating Rho-initiated pathways through plexin-B. Cancer Res 2004; 64(15):5212-5224.
46. Conrotto P, Valdembri D, Corso S et al. Sema4D induces angiogenesis through Met recruitment by Plexin B1. Blood 2005; 105(11):4321-4329.
47. Behar O, Golden JA, Mashimo H et al. Semaphorin III is needed for normal patterning and growth of nerves, bones and heart. Nature 1996; 383(6600):525-528.
48. Toyofuku T, Zhang H, Kumanogoh A et al. Dual roles of Sema6D in cardiac morphogenesis through region-specific association of its receptor, Plexin-A1, with off-track and vascular endothelial growth factor receptor type 2. Genes Dev 2004; 18(4):435-447.
49. Kikutani H. Semaphorin signaling during cardiac development. In: Pasterkamp RJ, ed. Semaphorins: Receptor and Intracellular Signaling Mechanisms. Georgetown: Landes Bioscience, 2006.
50. Ito T, Kagoshima M, Sasaki Y et al. Repulsive axon guidance molecule Sema3A inhibits branching morphogenesis of fetal mouse lung. Mech Dev 2000; 97(1-2):35-45.
51. Kagoshima M, Ito T. Diverse gene expression and function of semaphorins in developing lung: positive and negative regulatory roles of semaphorins in lung branching morphogenesis. Genes Cells 2001; 6(6):559-571.
52. Roy PJ, Zheng H, Warren CE et al. mab-20 encodes Semaphorin-2a and is required to prevent ectopic cell contacts during epidermal morphogenesis in Caenorhabditis elegans. Development 2000 Feb ;127 (4):755 -67 2000; 127(4):755-767.
53. Ginzburg VE, Roy PJ, Culotti JG. Semaphorin 1a and semaphorin 1b are required for correct epidermal cell positioning and adhesion during morphogenesis in C. elegans. Development 2002; 129(9):2065-2078.
54. Fujii T, Nakao F, Shibata Y et al. Caenorhabditis elegans PlexinA, PLX-1, interacts with transmembrane semaphorins and regulates epidermal morphogenesis. Development 2002; 129(9):2053-2063.
55. Liu Z, Fujii T, Nukazuka A et al. C. elegans PlexinA PLX-1 mediates a cell contact-dependent stop signal in vulval precursor cells. Dev Biol 2005; 282(1):138-151.
56. Dalpe G, Brown L, Culotti JG. Vulva morphogenesis involves attraction of plexin 1-expressing primordial vulva cells to semaphorin 1a sequentially expressed at the vulva midline. Development 2005; 132(6):1387-1400.
57. Holmes S, Downs AM, Fosberry A et al. Sema7A is a potent monocyte stimulator. Scand J Immunol 2002; 56(3):270-275.

58. Chabbert-de Ponnat I, Marie-Cardine A, Pasterkamp RJ et al. Soluble CD100 functions on human monocytes and immature dendritic cells require plexin C1 and plexin B1, respectively. Int Immunol 2005; 17(4):439-447.
59. Walzer T, Galibert L, De Smedt T. Dendritic cell function in mice lacking Plexin C1. Int Immunol 2005; 17(7):943-950.
60. Shi W, Kumanogoh A, Watanabe C et al. The class IV semaphorin CD100 plays nonredundant roles in the immune system: defective B and T cell activation in CD100-deficient mice. Immunity 2000; 13(5):633-642.
61. Kumanogoh A, Shikina T, Suzuki K et al. Nonredundant roles of Sema4A in the immune system: defective T cell priming and Th1/Th2 regulation in Sema4A-deficient mice. Immunity 2005; 22(3):305-316.
62. Wong AW, Brickey WJ, Taxman DJ et al. CIITA-regulated plexin-A1 affects T-cell-dendritic cell interactions. Nat Immunol 2003; 4(9):891-898.
63. Potiron V, Nasarre P, Roche J, Healy C, Boumsell L. Semaphorin signaling in the immune system. In: Pasterkamp RJ, ed. Semaphorins: Receptor and Intracellular Signaling Mechanisms. Georgetown: Landes Bioscience, 2006.
64. Christensen CR, Klingelhofer J, Tarabykina S et al. Transcription of a novel mouse semaphorin gene, M-semaH, correlates with the metastatic ability of mouse tumor cell lines. Cancer Res 1998; 58(6):1238-1244.
65. Kuroki T, Trapasso F, Yendamuri S et al. Allelic loss on chromosome 3p21.3 and promoter hypermethylation of semaphorin 3B in non-small cell lung cancer. Cancer Res 2003; 63(12):3352-3355.
66. Miao HQ, Lee P, Lin H et al. Neuropilin-1 expression by tumor cells promotes tumor angiogenesis and progression. FASEB J 2000; 14(15):2532-2539.
67. Christensen C, Ambartsumian N, Gilestro G et al. Proteolytic processing converts the repelling signal Sema3E into an inducer of invasive growth and lung metastasis. Cancer Res 2005; 65(14):6167-6177.
68. Barberis D, Artigiani S, Casazza A et al. Plexin signaling hampers integrin-based adhesion, leading to Rho-kinase independent cell rounding, and inhibiting lamellipodia extension and cell motility. FASEB J 2004; 18(3):592-594.
69. Conrotto P, Corso S, Gamberini S et al. Interplay between scatter factor receptors and B plexins controls invasive growth. Oncogene 2004; 23(30):5131-5137.
70. Giordano S, Corso S, Conrotto P et al. The semaphorin 4D receptor controls invasive growth by coupling with Met. Nat Cell Biol 2002; 4(9):720-724.
71. Kusy S, Nasarre P, Chan D et al. Selective suppression of in vivo tumorigenicity by semaphorin SEMA3F in lung cancer cells. Neoplasia 2005; 7(5):457-465.
72. Maestrini E, Tamagnone T, Longati P et al. A family of transmembrane proteins with homology to the MET-hepatocyte growth factor receptor. Proc Natl Acad Sci USA 1996; 93:674-678.
73. Rohm B, Rahim B, Kleiber B et al. The semaphorin 3A receptor may directly regulate the activity of small GTPases. FEBS Lett 2000; 486(1):68-72.
74. Pasterkamp RJ, Kolodkin AL. Semaphorin junction: making tracks toward neural connectivity. Curr Opin Neurobiol 2003; 13(1):79-89.
75. Puschel AW. GTPases in semaphorin signaling. In: Pasterkamp RJ, ed. Semaphorins: Receptor and Intracellular Signaling Mechanisms. Georgetown: Landes Bioscience, 2006.
76. Driessens MH, Hu H, Nobes CD et al. Plexin-B semaphorin receptors interact directly with active Rac and regulate the actin cytoskeleton by activating Rho. Curr Biol 2001; 11(5):339-344.
77. Vikis HG, Li W, He Z et al. The semaphorin receptor plexin-B1 specifically interacts with active Rac in a ligand-dependent manner. Proc Natl Acad Sci USA 2000; 97(23):12457-12462.
78. Turner LJ, Nicholls S, Hall A. The activity of the plexin-A1 receptor is regulated by Rac. J Biol Chem 2004; 279(32):33199-33205.
79. Jin Z, Strittmatter SM. Rac1 mediates collapsin-1-induced growth cone collapse. J Neurosci 1997; 17(16):6256-6263.
80. Kuhn TB, Brown MD, Wilcox CL et al. Myelin and collapsin-1 induce motor neuron growth cone collapse through different pathways: inhibition of collapse by opposing mutants of rac1. J Neurosci 1999; 19(6):1965-1975.
81. Vastrik I, Eickholt BJ, Walsh FS et al. Sema3A-induced growth-cone collapse is mediated by Rac1 amino acids 17- 32. Curr Biol 1999; 9(18):991-998.
82. Jurney WM, Gallo G, Letourneau PC et al. Rac1-mediated endocytosis during ephrin-A2- and semaphorin 3A-induced growth cone collapse. J Neurosci 2002; 22(14):6019-6028.

83. Hu H, Marton TF, Goodman CS. Plexin B mediates axon guidance in Drosophila by simultaneously inhibiting active Rac and enhancing RhoA signaling. Neuron 2001; 32(1):39-51.
84. Vikis HG, Li W, Guan KL. The plexin-B1/Rac interaction inhibits PAK activation and enhances Sema4D ligand binding. Genes Dev 2002; 16(7):836-845.
85. Artigiani S, Barberis D, Fazzari P et al. Functional regulation of semaphorin receptors by proprotein convertases. J Biol Chem 2003; 278(12):10094-10101.
86. Zanata SM, Hovatta I, Rohm B et al. Antagonistic effects of Rnd1 and RhoD GTPases regulate receptor activity in Semaphorin 3A-induced cytoskeletal collapse. J Neurosci 2002; 22(2):471-477.
87. Oinuma I, Katoh H, Harada A et al. Direct interaction of Rnd1 with Plexin-B1 regulates PDZ-RhoGEF-mediated Rho activation by Plexin-B1 and induces cell contraction in COS-7 cells. J Biol Chem 2003; 278(28):25671-25677.
88. Castellani V, De Angelis E, Kenwrick S et al. Cis and trans interactions of L1 with neuropilin-1 control axonal responses to semaphorin 3A. EMBO J 2002; 21(23):6348-6357.
89. Kantor DB, Chivatakarn O, Peer KL et al. Semaphorin 5A is a bifunctional axon guidance cue regulated by heparan and chondroitin sulfate proteoglycans. Neuron 2004; 44(6):961-975.
90. Song H, Ming G, He Z et al. Conversion of neuronal growth cone responses from repulsion to attraction by cyclic nucleotides [see comments]. Science 1998; 281(5382):1515-1518.
91. Basile JR, Afkhami T, Gutkind JS. Semaphorin 4D/plexin-B1 induces endothelial cell migration through the activation of PYK2, Src, and the phosphatidylinositol 3-kinase-Akt pathway. Mol Cell Biol 2005; 25(16):6889-6898.
92. Zhang Z, Vuori K, Wang H et al. Integrin activation by R-ras. Cell 1996; 85(1):61-69.
93. Wozniak MA, Kwong L, Chodniewicz D et al. R-Ras controls membrane protrusion and cell migration through the spatial regulation of Rac and Rho. Mol Biol Cell 2005; 16(1):84-96.
94. Kinbara K, Goldfinger LE, Hansen M et al. Ras GTPases: integrins' friends or foes? Nat Rev Mol Cell Biol 2003; 4(10):767-776.
95. Marte BM, Rodriguez-Viciana P, Wennstrom S et al. R-Ras can activate the phosphoinositide 3-kinase but not the MAP kinase arm of the Ras effector pathways. Curr Biol 1997; 7(1):63-70.
96. Berrier AL, Mastrangelo AM, Downward J et al. Activated R-ras, Rac1, PI 3-kinase and PKCepsilon can each restore cell spreading inhibited by isolated integrin beta1 cytoplasmic domains. J Cell Biol 2000; 151(7):1549-1560.
97. Jeong HW, Nam JO, Kim IS. The COOH-terminal end of R-Ras alters the motility and morphology of breast epithelial cells through Rho/Rho-kinase. Cancer Res 2005; 65(2):507-515.
98. Arthur WT, Burridge K. RhoA inactivation by p190RhoGAP regulates cell spreading and migration by promoting membrane protrusion and polarity. Mol Biol Cell 2001; 12(9):2711-2720.
99. Barberis D, Casazza A, Sordella R et al. p190 Rho-GTPase activating protein associates with plexins and it is required for semaphorin signaling. J Cell Sci 2005; 118(Pt 20):4689-4700.
100. Ridley AJ, Hall A. The small GTP-binding protein rho regulates the assembly of focal adhesions and actin stress fibers in response to growth factors. Cell 1992; 70(3):389-399.
101. Raftopoulou M, Hall A. Cell migration: Rho GTPases lead the way. Dev Biol 2004; 265(1):23-32.
102. Komatsu M, Ruoslahti E. R-Ras is a global regulator of vascular regeneration that suppresses intimal hyperplasia and tumor angiogenesis. Nat Med 2005.
103. Driessens MH, Olivo C, Nagata K et al. B plexins activate Rho through PDZ-RhoGEF. FEBS Lett 2002; 529(2-3):168-172.
104. Hirotani M, Ohoka Y, Yamamoto T et al. Interaction of plexin-B1 with PDZ domain-containing Rho guanine nucleotide exchange factors. Biochem Biophys Res Commun 2002; 297(1):32-37.
105. Perrot V, Vazquez-Prado J, Gutkind JS. Plexin B Regulates Rho through the Guanine Nucleotide Exchange Factors Leukemia-associated Rho GEF (LARG) and PDZ-RhoGEF. J Biol Chem 2002; 277(45):43115-43120.
106. Swiercz JM, Kuner R, Behrens J et al. Plexin-B1 directly interacts with PDZ-RhoGEF/LARG to regulate RhoA and growth cone morphology. Neuron 2002; 35(1):51-63.
107. Aurandt J, Vikis HG, Gutkind JS et al. The semaphorin receptor plexin-B1 signals through a direct interaction with the Rho-specific nucleotide exchange factor, LARG. Proc Natl Acad Sci USA 2002; 99(19):12085-12090.
108. Swiercz JM, Kuner R, Offermanns S. Plexin-B1/RhoGEF-mediated RhoA activation involves the receptor tyrosine kinase ErbB-2. J Cell Biol 2004; 165(6):869-880.
109. Banerjee J, Wedegaertner PB. Identification of a novel sequence in PDZ-RhoGEF that mediates interaction with the actin cytoskeleton. Mol Biol Cell 2004; 15(4):1760-1775.
110. Toyofuku T, Yoshida J, Sugimoto T et al. FARP2 triggers signals for Sema3A-mediated axonal repulsion. Nat Neurosci 2005; 8(12):1712-1719.

111. van Horck FP, Lavazais E, Eickholt BJ et al. Essential role of type I(alpha) phosphatidylinositol 4-phosphate 5-kinase in neurite remodeling. Curr Biol 2002; 12(3):241-245.
112. Eickholt BJ. Protein kinases in semaphorin signaling. In: Pasterkamp RJ, ed. Semaphorins: Receptor and Intracellular Signaling Mechanisms. Georgetown: Landes Bioscience, 2006.
113. Artigiani S, Conrotto P, Fazzari P et al. Plexin-B3 is a functional receptor for semaphorin 5A. EMBO Rep 2004; 5(7):710-714.
114. Winberg ML, Tamagnone L, Bai J et al. The transmembrane protein Off-track associates with Plexins and functions downstream of Semaphorin signaling during axon guidance. Neuron 2001; 32(1):53-62.
115. Atwal JK, Singh KK, Tessier-Lavigne M et al. Semaphorin 3F antagonizes neurotrophin-induced phosphatidylinositol 3-kinase and mitogen-activated protein kinase kinase signaling: a mechanism for growth cone collapse. J Neurosci 2003; 23(20):7602-7609.
116. Takagi S, Kasuya Y, Shimizu M et al. Expression of a cell adhesion molecule, neuropilin, in the developing chick nervous system. Dev Biol 1995; 170(1):207-222.
117. Ohta K, Mizutani A, Kawakami A et al. Plexin: a novel neuronal cell surface molecule that mediates cell adhesion via a homophilic binding mechanism in the presence of calcium ions. Neuron 1995; 14(6):1189-1199.
118. Hartwig C, Veske A, Krejcova S et al. Plexin B3 promotes neurite outgrowth, interacts homophilically, and interacts with Rin. BMC Neurosci 2005; 6:53.
119. Shimizu M, Murakami Y, Suto F et al. Determination of cell adhesion sites of neuropilin-1. J Cell Biol 2000; 148(6):1283-1293.
120. Carmeliet P, Tessier-Lavigne M. Common mechanisms of nerve and blood vessel wiring. Nature 2005; 436(7048):193-200.
121. Kurschat P, Bielenberg D, Rossignol M et al. Neuron restrictive silencer factor NRSF/REST is a transcriptional repressor of neuropilin-1 and diminishes the ability of semaphorin 3A to inhibit keratinocyte migration. J Biol Chem 2005; 281(5):2721-2729.
122. Mikule K, Gatlin JC, de la Houssaye BA et al. Growth cone collapse induced by semaphorin 3A requires 12/15-lipoxygenase. J Neurosci 2002; 22(12):4932-4941.
123. Kashiwagi H, Shiraga M, Kato H et al. Negative regulation of platelet function by a secreted cell repulsive protein, semaphorin 3A. Blood 2005; 106(3):913-921.
124. Delaire S, Billard C, Tordjman R et al. Biological activity of soluble CD100. II. Soluble CD100, similarly to H-SemaIII, inhibits immune cell migration. J Immunol 2001; 166(7):4348-4354.
125. Bachelder RE, Lipscomb EA, Lin X et al. Competing autocrine pathways involving alternative neuropilin-1 ligands regulate chemotaxis of carcinoma cells. Cancer Res 2003; 63(17):5230-5233.
126. Nasarre P, Constantin B, Rouhaud L et al. Semaphorin SEMA3F and VEGF have opposing effects on cell attachment and spreading. Neoplasia 2003; 5(1):83-92.
127. Nasarre P, Kusy S, Constantin B et al. Semaphorin SEMA3F has a repulsing activity on breast cancer cells and inhibits E-cadherin-mediated cell adhesion. Neoplasia 2005; 7(2):180-189.
128. Delorme G, Saltel F, Bonnelye E et al. Expression and function of semaphorin 7A in bone cells. Biol Cell 2005; 97(7):589-597.

CHAPTER 9

Semaphorin Signaling during Cardiac Development

Toshihiko Toyofuku and Hitoshi Kikutani*

A number of semaphorins have been shown to play crucial roles as axon guidance cues in the wiring of the nervous system, including axon fasciculation, branching, and target selection. However, increasing evidence has also attested to the significance of semaphorins in the development of other organ systems, including the cardiovascular system. Targeted disruption of certain semaphorins or their receptors has been shown to result in various defects in the vascular system. Furthermore, several studies have suggested that some semaphorins may contribute to the development of the cardiovascular system by controlling the migration of endothelial cells, cardiac myocytes, or their precursors. In this review, we will discuss how semaphorin signals are involved in regulation of cardiac cells and cardiac morphogenesis.

Cardiac Morphogenesis: An Overview

The heart is one of the first mesodermal tissues to differentiate just after gastrulation in the vertebrate embryo.[1,2] Cells that migrate anterior and lateral to the primitive streak in early gastrulation contribute to heart tissues (Fig. 1A). Soon after their specification, precursors of cardiac cells converge along the ventral midline of the embryo to form a linear heart tube composed of myocardial and endocardial layers separated by an extracellular matrix (Fig. 1B). In all vertebrates, the linear heart tube undergoes rightward looping, which is essential for proper orientation of the right and left ventricles, and for alignment of the heart chambers with the vasculature (Fig. 1C). The direction of cardiac looping is determined by an asymmetric axial signaling system involving Nodal, Lefty, and Pitx2, which also affects the position of the lungs, liver, spleen, and gut.[3] The linear heart tube becomes segmentally patterned along the cranial-caudal axis into precursors of the aortic sac, conotruncus, pulmonary and systemic ventricles, and atria. Upon rightward looping, the cranial (conotruncus) and the caudal portions of the cardiac tube juxtapose dorsally and fuse to form a single outflow tract (truncus arteriosus), and the middle portion of cardiac tube expands to form a single ventricle (Fig. 1D). In the developing ventricle, interventricular septation separates the right and left ventricles, each of which differs in its morphological and contractile properties. The left ventricle is composed of distinct outer (compact) and inner (trabecular) layers. Trabeculae, finger-like projections comprised of myocardial cells, are necessary to support the increasing hemodynamic load and to supply nutrients from inside without blood vessels during early embryonic heart development.[4]

*Corresponding Author: Hitoshi Kikutani—Department of Molecular Immunology, Research Institute for Microbial Diseases, Osaka University, 3-1 Yamada-oka, Suita, Osaka 565-0871, Japan. CREST, Japan Science and Technology Corporation, Japan. Email: kikutani@ragtime.biken.osaka-u.ac.jp

Semaphorins: Receptor and Intracellular Signaling Mechanisms, edited by R. Jeroen Pasterkamp. ©2007 Landes Bioscience and Springer Science+Business Media.

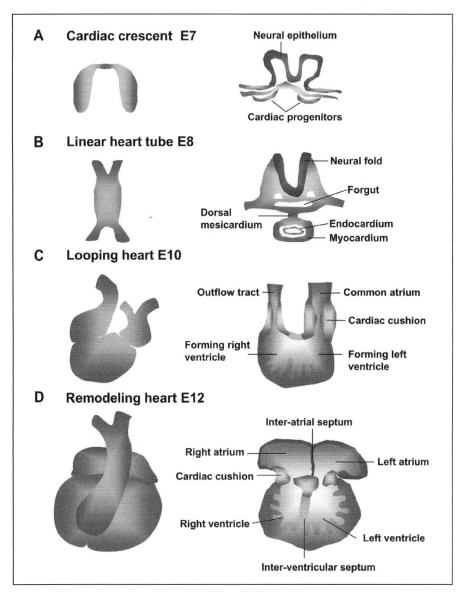

Figure 1. The main transitions in early heart development. The whole isolated heart is shown on the left, whereas a representative section is presented on the right. Staging in days of embryonic development is based on mouse development. A) Cardiac progenitors are first recognizable as a crescent-shaped epithelium (the cardiac crescent). B) Heart progenitors move ventrally to form the linear heart tube, which is composed of an endothelial lining that is surrounded by a myocardial epithelium. C) The linear heart tube undergoes a complex progression termed cardiac looping, including endocardial cushion formation in the atrioventricular canal and outflow tract, and trabecular formation in the ventricle. D) During the remodeling phase of heart development, division of chambers by septation is completed. The chambers and vessels are now aligned as in the adult heart and become fully integrated.

Appropriate placement and function of cardiac valves is essential for division of the chambers and for unidirectional flow of blood through the heart. Septation of the cardiac tube into distinct chambers is achieved through regional swellings of extracellular matix with proliferating mesenchymal cells, known as cardiac cushions, that will form the atrioventricular and ventriculoarterial valves. The formation of cardiac cushions is a complex event characterized by the endothelial-to-mesenchymal transformation of a subset of endothelial cells in the cushion-forming region, where they subsequently proliferate and complete their differentiation into mesenchymal cells.[5]

The separation of the pulmonary and systemic circulations into two parallel circuits is established by the separation of the truncus arteriosus into the aorta and pulmonary artery, and in the formation of the conotruncal portion of the ventricular septum. This is accomplished in part by cardiac neural crest cells.[6,7] Neural crest cells, which originate from neuroepithelial cells at the dorsal edge of neural tube, undergo epithelial-to-mesenchymal transformation so that they migrate to form part of the mesenchyme of the outflow tract. After septation, the vessels rotate in a twisting fashion to achieve their connections with the right and left ventricles. Thus, the mechanism that regulates the positioning of myocardial cells in the dynamic remodeling of the cardiac tube may be necessary for the proper alignment and structure of each chamber in that structure. It has been recently shown that some semaphorins play roles in the positioning of cardiac cells during cardiac development by regulating their migration.

Sema6D-Plexin-A1 Axis in Cardiac Morphogenesis

One of the best characterized semaphorins in cardiac development is Sema6D, a member of the class VI transmembrane-type semaphorin subfamily.[8] The expression of Sema6D is first detected in the cardiac crescent and neural fold of E9 mouse embryo. Sema6D mRNA was observed throughout the entire heart, including the conotruncal (CT) segment, the atrioventricular segment, and the ventricular myocardium at E10.5. The expression of Sema6D is higher in myocardial cells than in endocardial cells. In chick embryonic heart, Sema6D exhibits a similar expression pattern.

The role of Sema6D in the developing heart has been revealed by a series of studies using the chick embryo system. Inoculation of transfected cells that release a large amount of soluble Sema6D into cultured chick embryos at Hamberger and Hamilton (HH) stage 9 results in enhanced looping of the cardiac tube and enlargement of the ventricular region. In ovo inoculation of Sema6D producing cells or recombinant soluble Sema6D into HH stage 29 embryos also results in expansion of the ventricular cavity with a thin myocardial layer and an enlarged endocardial cushion. In contrast, RNAi-mediated knockdown of Sema6D inhibits looping of the cardiac tube. In the developing heart, Sema6D signals are largely mediated by Plexin-A1, which is also expressed in the embryonic heart. Indeed, RNAi-mediated knockdown of Plexin-A1 or expression of truncated Plexin-A1 results in decreased ventricular size. Therefore, the Sema6D-Plexin-A1 axis is critically involved in the dynamic remodeling of the cardiac tube and formation of the ventricle and endocardial cushion.

Sema6D Differentially Regulates Migration of Endothelial Cells in Distinct Regions of Cardiac Tube

Sema6D exerts distinct biological activities on endothelial cells in different regions of the cardiac tube. For instance, Sema6D inhibits migration of outgrowing cells from the ventricular segment. On the other hand, Sema6D promotes migration of outgrowing cells from the conotruncal and atrioventricular valve segments, both of which fuse to form the endocardial cushion later. These biological activities of Sema6D appear to be mediated by Plexin-A1, because they are abrogated by RNAi-mediated knockdown of Plexin-A1 or expression of truncated Plexin-A1 in endothelial cells from the ventricle as well as the conotruncal segments. How are two distinct biological activities mediated through the same ligand binding receptor? In the nervous system of *Drosophila*, Plexin-A, a homologue of mammalian Plexin-A family members, forms a complex with off-track (OTK), to transduce the repulsive signaling of Sema-1a, a homologue of mammalian class VI semaphorins.[9] Like *Drosophila* Plexin-A, Plexin-A1 forms a complex with a vertebrate homologue of OTK in the endothelial cells of the ventricular region of the cardiac tube. The migration-inhibitory activity of Sema6D is suppressed by RNAi-mediated knockdown of OTK in ventricular endothelial cells. OTK is a member of the membrane-type tyrosine kinase family, although its kinase activity is lost because of the absence of critical residues in the kinase domain (so-called kinase dead). Notably, mice lacking the mammalian homologue to OTK exhibit several defects, including failure of tube closure, which are known to be associated with defective planar cell polarity.[10] Thus, Sema6D functions as a positional cue for migrating cells through the Plexin-A1/OTK receptor complex so as to regulate the shape and rotation of the cardiac ventricle (Fig. 2).

On the other hand, Plexin-A1 forms a complex with VEGF receptor type 2 (VEGFR2) in endothelial cells of conotruncal segments. The Sema6D-induced migration of endothelial cells is suppressed not only by RNAi against Plexin-A1, but also by RNAi against VEGFR2. A narrow range of VEGF levels is critical for cardiac cushion formation, because either induction or suppression of VEGF induces similar defects in cardiac cushion formation.[11,12] The VEGF-mediated tyrosine phosphorylation of VEGFR2, an initial step in the VEGF signaling pathway, is enhanced by Sema6D. Thus, Sema6D functions to regulate cardiac-cushion formation by modifying the signaling of VEGFR2 through Plexin-A1 on endothelial cells in the endocardiac cushion-forming region (Fig. 2). Indeed, overexpression of Sema6D results in enlargement of endocardiac cushion. Thus, the differential association of Plexin-A1 with additional receptor components enables Sema6D to exert distinct biological activities in adjacent regions, which is critical for complex cardiac morphogenesis.

Reverse Signaling of Sema6D in the Cardiac Ventricle

The developing ventricular wall is composed of the outer myocardial layer (compact layer) and trabeculae. Myocardial cells in the former express both Sema6D and Plexin-A1, and cells in the latter express Sema6D but not Plexin-A1.[13] Knockdown of either Sema6D or Plexin-A1 leads to the generation of a small, thin ventricular compact layer and to defective trabeculation. Ectopic expression of the Plexin-A1 extracellular domain alone can rescue the defective trabeculation induced by the suppression of Plexin-A1 but not that induced by suppression of Sema6D, indicating a role for Sema6D cytoplasmic signaling in trabeculation (Fig. 3A). The Sema6D cytoplasmic region can associate with two molecules; Abl kinase and Enabled (Ena), a member of the Ena/VASP family. Abl kinase and Ena are known to play opposing roles in the downstream regulation of *Drosophila* Robo, an axonal guidance receptor for Slit.[14] Ena has also been implicated in reverse signaling by *Drosophila* Sema1a. Upon binding to Plexin-A1, Abl kinase is recruited to the cytoplasmic tail of Sema6D and activated, which results in phosphorylation of Ena and its dissociation from Sema6D. In fate-mapping studies, myocardial cells carrying defects in reverse signaling by Sema6D arrest in the compact layer, whereas expression of constitutively active Abl kinase enhances the migration of cells from the compact layer to the trabeculae. Thus, Sema6D acts through its cytoplasmic domain as a reverse signal to regulate trabeculation.

Semaphorin Signaling in Vascular Connections to the Heart

Neuropilins, a receptor for class-III semaphorins, are widely expressed in the developing vasculature, and mice lacking neuropilin-1 or neuropilin-1 and -2 exhibit branching and

Semaphorin Signaling during Cardiac Development

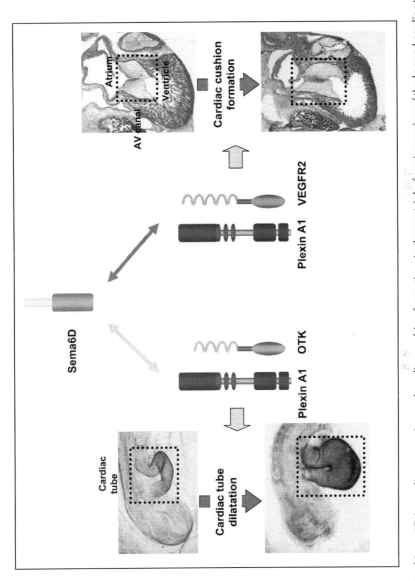

Figure 2. The role of Sema6D in cardiac expansion and cardiac cushion formation. In the ventricle-forming region of the looped cardiac tube, Plexin-A1 and OTK form a receptor complex and mediate Sema6D-induced suppression of endothelial cell motility, leading to cardiac expansion. In the cardiac cushion-forming region, Plexin-A1 and VEGFR2 form a receptor complex and mediate Sema6D-induced enhancement of endothelial cell motility, leading to cardiac-cushion formation.

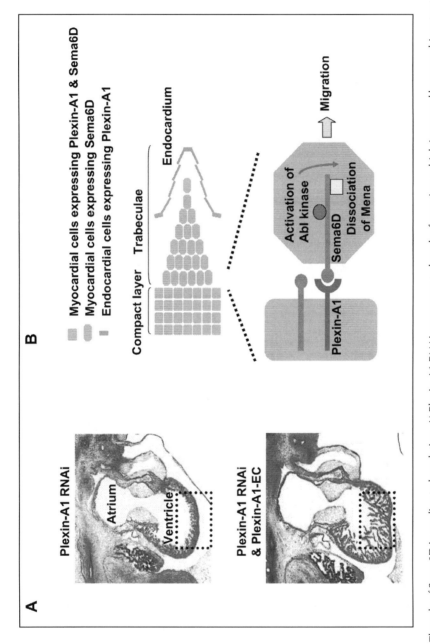

Figure 3. The role of Sema6D in cardiac trabeculation. A) Plexin-A1 RNAi suppresses trabecular formation, which is rescued by recombinant extracellular domain of plexin-A1 (B) The role of Sema6D in cardiac trabeculation through its cytoplasmic region. Plexin-A1 binding to Sema6D activates Abl kinase, and activated Abl kinase phosphorylates the associated Mena, resulting in its dissociation. This process allows myocardial cells to migrate into trabeculae.

remodeling defects, improper routing and connections, and ectopic termination of vessels.[15] Sema3A inhibits formation of endothelial lamellipodia and vessels.[16-18] However, neuropilins are also receptors for a specific VEGF isoform (VEGF165) and modulate the activity of VEGF receptors;[19] moreover, VEGF165 competes with Sema3A for binding to neuropilins.[16] Thus, the vascular effects of neuropilin-1 may reflect loss of VEGF rather than Sema3A signaling. Mice expressing a variant of neuropilin-1 that are only capable of binding to VEGF, but not to semaphorins, do not exhibit vascular defects,[20] indicating that neuropilin-1 plays a major role in vascular patterning as a VEGF coreceptor.

Mice lacking Sema3C exhibit persistent truncus arteriosus.[21] Plexin-A2, which is usually detected in cardiac neural crest, is patterned abnormally in several mutant mouse lines with congenital heart disease, including those lacking the secreted signaling molecule Sema3C.[22] It is noteworthy that Sema3C shows an attractive effect on cortical neurons through neuropilins-1.[23] Together with these findings, the complementary expression pattern of Sema3C and Pexin-A2 raises the possibility that cardiac neural crest cells navigate from the neural tube to the outflow tract of the cardiac tube by attractive signals involving the Sema3C-Plexin-A2 axis.

Signals of Sema3A in Endothelial Cell Migration

As described above, class III semaphorins such as Sema3C may be involved in cardiac morphogenesis. However, it remains largely unclear how class III semaphorins regulate migration of cardiac cells. A recent report showing the involvement of Sema3A in the regulation of vascular endothelial cells may provide a clue to this issue. Serini et al[18] showed that Sema3A suppresses adhesion of endothelial cells, which may contribute to the regulation of endothelial cell migration during vasculogenesis. In fact, Sema3A inhibits the function of $\alpha v\beta 3$ and $\alpha v\beta 5$ integrins in endothelial cells, although the molecular mechanism underlying this regulation is not known.

We have recently shown that the FERM domain-containing guanine nucleotide exchange factor (GEF) FARP2 functions as an immediate downstream signal transducer of the Plexin-A1-neuropilin-1 receptor complex in Sema3A-mediated repulsion of axons[24] (Fig. 4). Sema3A-induced dissociation of FARP2 from Plexin-A1 and activation of its Rac-GEF activity triggers a series of biochemical events including Rac activation and the binding of Rnd1, a member of the Rho GTPase family, to Plexin-A1.[25] This binding stimulates the GAP activity of Plexin-A1 for R-Ras, a member of the Ras GTPase family. Thus, the downregulation of R-Ras leads to cytoskeletal disassembly, which is critical for Plexin-A1-mediated growth-cone collapse.[26] In parallel with this event, dissolved FARP2 competes with an isoform of type-1 phosphatidylinositol phosphate kinase, PIPKIγ661, for the FERM domain of talin. PIP_2, catalyzed by PIPKIγ661 in association with talin, is important for the stability of integrin-mediated focal adhesion,[27,28] and the inhibition of PIPKIγ661 kinase activity by binding to FARP2 downregulates integrin function. Such a mechanism for class III semaphorin-mediated regulation of cell adhesion may also be involved in control of migration of precursors of cardiac cells during embryonic development.

Summary and Perspectives

As discussed in this review, the patterning and morphogenesis of the heart can be described in some detail, but the relevant molecular details are not completely understood. To further comprehend heart development, we must completely define its basic units, then seek the points of integration between various molecular systems affecting the developing heart tube. The recent discovery that guidance molecules such as semaphorins regulate the dynamic remodeling of the cardiac tube should provide insights into the molecular mechanisms underlying cardiac morphogenesis.

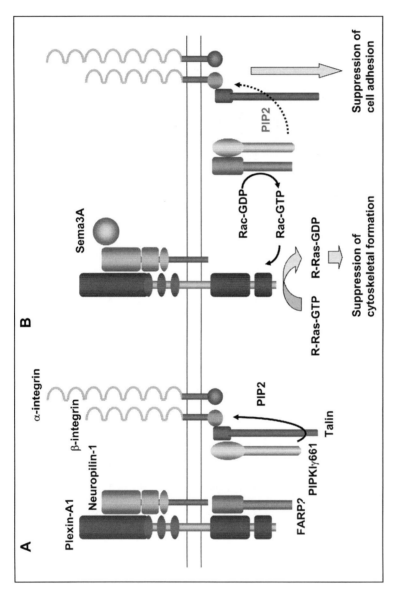

Figure 4. A schematic diagram depicting the roles of FARP2 in the initial steps of Sema3A-Plexin-A1 signaling. Sema3A binding to the receptor complex comprising Neuropilin-1 as the ligand-binding subunit and Plexin-A1 as the signal-transducing subunit triggers the dissociation of FARP2. Released FARP2 has two major roles in the downstream signaling of Plexin-A1. First, the RacGEF activity of FARP2 is turned on. This activity is essential for subsequent Rnd1 recruitment to Plexin-A1 and activation of Plexin-A1 downstream signaling events such as activation of R-Ras GAP activity of Plexin-A1 and downregulation of R-Ras. Second, released FARP2 binds to PIPKIγ661 and inhibits its PIPKIγ kinase activity, resulting in an inhibition of focal adhesion.

References

1. Fishman MC, Chien KR. Fashioning the vertebrate heart: Earliest embryonic decisions. Development 1997; 124(11):2099-2117.
2. Srivastava D, Olson EN. A genetic blueprint for cardiac development. Nature 2000; 407(6801):221-226.
3. Harvey RP. Patterning the vertebrate heart. Nat Rev Genet 2002; 3(7):544-556.
4. Sedmera D, Pexieder T, Vuillemin M et al. Developmental patterning of the myocardium. Anat Rec 2000; 258(4):319-337.
5. Eisenberg LM, Markwald RR. Molecular regulation of atrioventricular valvuloseptal morphogenesis. Circ Res 1995; 77(1):1-6.
6. Le Douarin NM, Creuzet S, Couly G et al. Neural crest cell plasticity and its limits. Development 2004; 131(19):4637-4650.
7. Stoller JZ, Epstein JA. Cardiac neural crest. Semin Cell Dev Biol 2005; 16(6):704-715, (Epub 2005 Jul 2027).
8. Toyofuku T, Zhang H, Kumanogoh A et al. Dual roles of Sema6D in cardiac morphogenesis through region-specific association of its receptor, Plexin-A1, with off-track and vascular endothelial growth factor receptor type 2. Genes Dev 2004; 18(4):435-447, (Epub 2004 Feb 2020).
9. Winberg ML, Tamagnone L, Bai J et al. The transmembrane protein Off-track associates with Plexins and functions downstream of Semaphorin signaling during axon guidance. Neuron 2001; 32(1):53-62.
10. Lu X, Borchers AG, Jolicoeur C et al. PTK7/CCK-4 is a novel regulator of planar cell polarity in vertebrates. Nature 2004; 430(6995):93-98.
11. Dor Y, Camenisch TD, Itin A et al. A novel role for VEGF in endocardial cushion formation and its potential contribution to congenital heart defects. Development 2001; 128(9):1531-1538.
12. Enciso JM, Gratzinger D, Camenisch TD et al. Elevated glucose inhibits VEGF-A-mediated endocardial cushion formation: Modulation by PECAM-1 and MMP-2. J Cell Biol 2003; 160(4):605-615.
13. Toyofuku T, Zhang H, Kumanogoh A et al. Guidance of myocardial patterning in cardiac development by Sema6D reverse signaling. Nat Cell Biol 2004; 6(12):1204-1211, (Epub).
14. Bashaw GJ, Kidd T, Murray D et al. Repulsive axon guidance: Abelson and Enabled play opposing roles downstream of the roundabout receptor. Cell 2000; 101(7):703-715.
15. Kawasaki T, Kitsukawa T, Bekku Y et al. A requirement for neuropilin-1 in embryonic vessel formation. Development 1999; 126(21):4895-4902.
16. Miao HQ, Soker S, Feiner L et al. Neuropilin-1 mediates collapsin-1/semaphorin III inhibition of endothelial cell motility: Functional competition of collapsin-1 and vascular endothelial growth factor-165. J Cell Biol 1999; 146(1):233-242.
17. Shoji W, Isogai S, Sato-Maeda M et al. Semaphorin3a1 regulates angioblast migration and vascular development in zebrafish embryos. Development 2003; 130(14):3227-3236.
18. Serini G, Valdembri D, Zanivan S et al. Class 3 semaphorins control vascular morphogenesis by inhibiting integrin function. Nature 2003; 424(6947):391-397.
19. Soker S, Takashima S, Miao HQ et al. Neuropilin-1 is expressed by endothelial and tumor cells as an isoform-specific receptor for vascular endothelial growth factor. Cell 1998; 92(6):735-745.
20. Gu C, Rodriguez ER, Reimert DV et al. Neuropilin-1 conveys semaphorin and VEGF signaling during neural and cardiovascular development. Dev Cell 2003; 5(1):45-57.
21. Feiner L, Webber AL, Brown CB et al. Targeted disruption of semaphorin 3C leads to persistent truncus arteriosus and aortic arch interruption. Development 2001; 128(16):3061-3070.
22. Brown CB, Feiner L, Lu MM et al. PlexinA2 and semaphorin signaling during cardiac neural crest development. Development 2001; 128(16):3071-3080.
23. Bagnard D, Thomasset N, Lohrum M et al. Spatial distributions of guidance molecules regulate chemorepulsion and chemoattraction of growth cones. J Neurosci 2000; 20(3):1030-1035.
24. Toyofuku T, Yoshida J, Sugimoto T et al. FARP2 triggers signals for Sema3A-mediated axonal repulsion. Nat Neurosci 2005; 8(12):1712-1719, (Epub 2005 Nov 1713).
25. Zanata SM, Hovatta I, Rohm B et al. Antagonistic effects of Rnd1 and RhoD GTPases regulate receptor activity in Semaphorin 3A-induced cytoskeletal collapse. J Neurosci 2002; 22(2):471-477.
26. Oinuma I, Ishikawa Y, Katoh H et al. The Semaphorin 4D receptor Plexin-B1 is a GTPase activating protein for R-Ras. Science 2004; 305(5685):862-865.
27. Ling K, Doughman RL, Firestone AJ et al. Type I gamma phosphatidylinositol phosphate kinase targets and regulates focal adhesions. Nature 2002; 420(6911):89-93.
28. Di Paolo G, Pellegrini L, Letinic K et al. Recruitment and regulation of phosphatidylinositol phosphate kinase type 1 gamma by the FERM domain of talin. Nature 2002; 420(6911):85-89.

CHAPTER 10

Semaphorin Signaling in Vascular and Tumor Biology

Gera Neufeld,* Tali Lange, Asya Varshavsky and Ofra Kessler

Abstract

The neuropilins were originally characterized as cell membrane receptors that bind axon guidance factors belonging to the class-3 semaphorin subfamily. To transduce semaphorin signals, they form complexes with members of the plexin receptor family in which neuropilins serve as the ligand binding components and the plexins as the signal transducing components. The neuropilins were subsequently found to double as receptors for specific heparin binding splice forms of vascular endothelial growth factor (VEGF), and to be expressed on endothelial cells. This finding suggested that semaphorins may function as modulators of angiogenesis. It was recently found that several types of semaphorins such as semaphorin-3F function as inhibitors of angiogenesis while others, most notably semaphorin-4D, function as angiogenic factors. Furthermore, semaphorins such as semaphorin-3F and semaphorin-3B have been characterized as tumor suppressors and have been found to exert direct effects upon tumor cells. In this chapter we cover recent developments in this rapidly developing field of research.

Receptors Belonging to the Neuropilin and Plexin Families and Their Semaphorin Ligands

The Semaphorins

The semaphorins were originally identified as axon guidance factors that induce localized collapse of neuronal growth cones, which is why they were initially named collapsins.[1] The semaphorin family consists of more than 30 genes divided into 8 classes, of which the first two classes are derived from invertebrates, classes 3-7 are the products of vertebrate semaphorins, and the 8th class contains viral semaphorins (Fig. 1). The semaphorins were often referred to by an array of confusing designations. This situation was clarified by the adoption of a unified nomenclature for the semaphorins.[2] The semaphorins are characterized by the presence of a ~500 amino-acids long sema domain that is located close to their N-termini. The sema domain is essential for semaphorin signaling and determines the specificity of binding.[3] The X-ray structures of semaphorin-3A (Sema3A) and semaphorin-4D (Sema4D) sema domains were analyzed at the atomic level revealing a conserved β-propeller structure.[3] Class-3 semaphorins are distinguished from other vertebrate semaphorins in being the only secreted semaphorins.

*Corresponding Author: Dr. Gera Neufeld—Cancer and Vascular Biology Research Center, Rappaport Research Institute in the Medical Sciences, The Bruce Rappaport Faculty of Medicine, Technion, Israel Institute of Technology, 1 Efron St., P. O. Box 9679, Haifa, 31096, Israel. Email: gera@tx.technion.ac.il

Semaphorins: Receptor and Intracellular Signaling Mechanisms, edited by R. Jeroen Pasterkamp.
©2007 Landes Bioscience and Springer Science+Business Media.

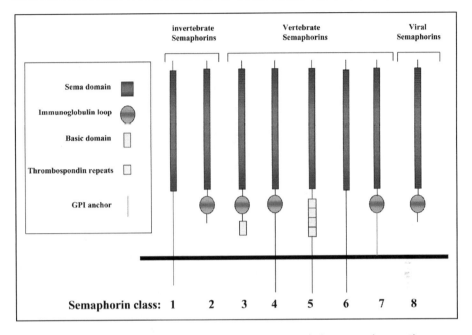

Figure 1. The semaphorin family. The different semaphorin subclasses are shown. Classes 3-7 contain vertebrate semaphorins. The two main semaphorin subclasses containing members reported to function as angiogenesis regulators are class-3 and class-4 semaphorins. Class-3 semaphorins are the only secreted vertebrate semaphorins. The subfamily contains seven known members. They are distinguished by a small basic domain and by an Ig-like domain in addition to the sema domain which is present in all semaphorins. Class-4 semaphorins are membrane anchored semaphorins containing an Ig loop-like domain. For more details see the text.

In addition class-3 semaphorins are distinguished by the presence of a basic domain in their C-terminus. Semaphorins belonging to the other classes are cell membrane bound proteins, and are distinguished by the presence of specific structural features such as the thrombospondin repeats found in class-5 semaphorins or by the GPI anchor found in class-7 semaphorins (Fig. 1). Active soluble semaphorins can be generated from some of these membrane anchored semaphorins by way of proteolytic processing.[4] Semaphorins repel growth cones of elongating axons by causing a localized collapse of the cytoskeleton in the growth cone thereby directing it in the opposite direction.[5] Semaphorins such as Sema3A were also found to function as inducers of apoptosis,[6,7] as modulators of cell migration,[8,9] and as factors that regulate cell adhesion.[10,11] When applied externally and nondirectionally, semaphorins induce a general collapse of the cytoskeleton in responsive cells which is manifested by cell contraction.[12] Most of the class-3 semaphorins bind to receptors belonging to the neuropilin family (Fig. 2). In-contrast, semaphorins belonging to classes 4-7 as well as Sema3E, do not bind to neuropilins. They bind directly to receptors of the plexin family (Fig. 3).[13,14]

Neuropilins

The neuropilin-1 (np1) receptor was initially identified as a neuronal cell surface antigen (A5) in xenopus embryos.[15] It was subsequently renamed neuropilin[16] and was found to function as a receptor for Sema3A. Almost simultaneously, a second member of the neuropilin receptor family, neuropilin-2 (np2), was identified and found to function as a receptor for another class-3 semaphorin, Sema3F.[17,18] The two neuropilins share a very similar domain structure although the overall homology between np1 and np2 is only 44% at the amino-acid

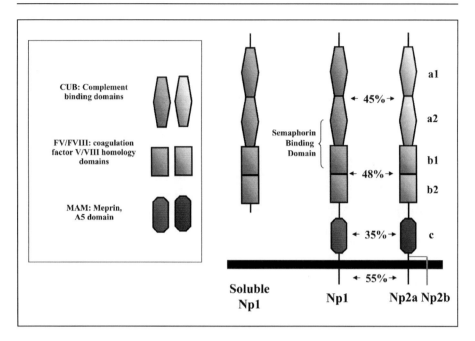

Figure 3. The neuropilin receptor family. The two members of the neuropilin family are membrane-anchored receptors containing very short intracellular domains. Interestingly, there exist two np2 splice forms in which the transmembrane and intracellular domains are completely different (np2a and np2b). The Sema3A binding domain of np1 is located between the a2 and b1 domains and overlaps partially with the $VEGF_{165}$ binding domain. The MAM domain is required for receptor dimerization and for interaction with other receptors. For more details see the text.

level.[19] Both neuropilins contain two complement binding (CUB) like domains (a1 and a2 domains), two coagulation factor V/VIII homology like domains (b1 and b2 domains), and a meprin (MAM) like domain thought to be important for neuropilin dimerization and possibly for the interaction of neuropilins with other membrane receptors (Fig. 2).[19,20] The binding site of Sema3A in np1 covers part of the second a-domain and part of the first b-domain.[21] The neuropilins are membrane spanning receptors but their intracellular domain is short and is believed to be too short for independent signal transduction. It was demonstrated that to transduce signals of class-3 semaphorins such as Sema3A and Sema3F the intracellular domain is not required.[22] It was subsequently found that neuropilins form complexes with members of the plexin family of receptors in which the neuropilin serves as the ligand binding part while the plexin transmits the signal through the membrane of the target cells.[12,14] Various class-3 semaphorins differentiate between the two neuropilin receptors. For example, it was found that Sema3A binds to np1 but not to np2, while Sema3F binds well to np2 but only with a much reduced affinity to np1.[17,19,20]

Plexins and Their Role in Semaphorin Signal Transduction

The vertebrate plexin family consists of nine genes grouped into four subfamilies. There are four A type plexins, three B type plexins, and single C and D type plexins.[23,24] Different plexins serve as direct binding receptors for different types of semaphorins. Thus, plexin-B1 and plexin-B2 are receptors for Sema4D,[14] plexin-B3 is a receptor for Sema5A,[25] plexin-A1 is a Sema6D receptor,[26] and plexin-D1 is a receptor for Sema3E[13] to name but a few examples. As mentioned before, most class-3 semaphorins do not bind directly to plexins but bind instead to

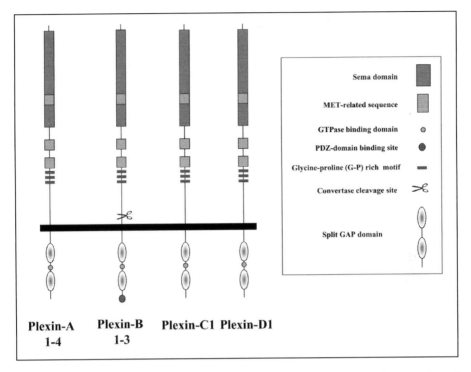

Figure 3. The plexin receptor family. There are currently nine known mammalian members of this family. They are grouped into 4 subfamilies. Members of the A, B and D subfamilies have been found to function as modulators of angiogenesis. All plexins contain a sema domain and MET-related sequences. The intracellular part contains tyrosine residues that can be phosphorylated but lack tyrosine-kinase activity and a split GAP domain.

neuropilins which associate with type-A plexins to form holo-receptors in which the plexins serve as the signal transducing element.[12,14] The plexins contain in their extracellular parts a sema domain which apparently functions as an inhibitory domain that prevents activation in the absence of ligand. In the case of plexin-A1, it was found that following the Sema3A binding to np1, the sema domain of plexin-A1 changes conformation enabling signal transduction while in the absence of the sema domain plexin-A1 was constitutively active.[27] The extracellular domain of the plexins contain in addition sequences homologous to sequences found in the Met subfamily of tyrosine-kinase receptors. These are designated Met related sequence (MRS) domains and glycine-proline (G-P) rich motifs (Fig. 3).[28] The plexins contain in their intracellular domain a split cytoplasmic SP (sex-plexin) domain (Also known as the C1 and C2 domains) which functions as a split GTPase activating (GAP) domain. This GAP like domain is conserved quite highly throughout the plexin family although it is unclear whether it is functional in all plexins.[29] In the absence of ligand this split GAP domain is inactive. Both plexin-B1 and plexin-A1 contain a GTPase binding domain located between the two halves of the split GAP domain which serves as a docking site for small GTPases such as Rac1, Rnd1 and R-Ras. In plexin-B1, it was found that the binding of Rnd1 activates the intrinsic GAP function of the plexin resulting in the inactivation of the small GTPase R-Ras. This in turn results in the inhibition of integrin function and localized cell detachment. A similar mechanism seems to control the activity of the plexin-A1 GAP domain.[29,30] In the case of plexin-A1, the binding of Rnd1 can be inhibited by the small GTPase RhoD and this in turn inhibits Sema3A induced signal transduction (Fig. 4).[31]

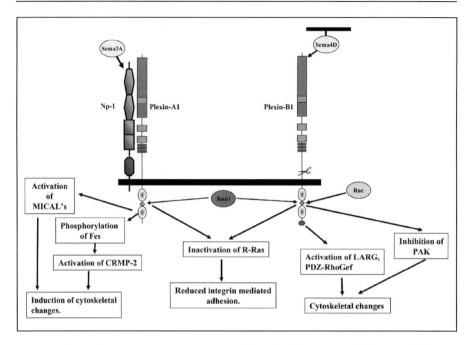

Figure 4. The main signaling pathways activated by the plexin-A1 and plexin-B1 following activation by their respective Sema3A and Sema4D ligands. Following the binding of Sema3A to np1, which is associated to plexin-A1 in the presence or absence of Sema3A, plexin-A1 is activated. Rnd1 binds to the GTPase binding site leading to activation of the intrinsic GAP domain, which leads to R-Ras inactivation and inhibition of integrin function. Simultaneously, other pathways involving activation of CRMP-2 and MICAL's lead to the reorganization of the actin and tubulin cytoskeleton. In the case of plexin-B1, activation occurs directly, without involvement of neuropilins. Inactivation of integrin function via R-Ras inactivation occurs similarly. Cytoskeletal changes are triggered by the activation of Rho GEF's via the PDZ binding domain of plexin-B1. For more details see the text.

Additional small GTPases and their corresponding Guanine-nucleotide exchange factors (GEFs) and GAPs participate in the transduction of plexin-mediated signals. It was shown that B-type plexins possess a postsynaptic density-95/Discs large/zona occludens-1 (PDZ) binding motif at the C-terminus through which GEFs such as PDZ-GEF and Leukemia-associated Rho-GEF (LARG) bind to B-type plexins.[32,33] The small GTPase Rho is activated following the binding of these Rho-GEFs to plexin-B1, leading to activation of Rho Kinase and initiation of a reorganization of the actin cytoskeleton in response to Sema4D.[34,35] The active form of the small GTPase Rac also binds to the GTPase binding domain of plexin-B1 and thereby enhances the binding of Sema4D to plexin-B1. To transduce signals, active Rac interacts with p21-activated kinase (PAK) (Fig. 4). Activated plexin-B1 competes with PAK by sequestering active Rac thereby inhibiting the activation of PAK which is one of the signal transducers that control actin polymerization.[36]

The intracellular domain of the plexins does not contain a tyrosine-kinase domain. However, intracellular tyrosine-kinases such as Fes/Fps bind to Sema3A activated plexin-A1 resulting in the phosphorylation of tyrosine residues in the intracellular domain of plexin-A1.[37] Activated Fes also phosphorylates the CRMP-2 and CRAM proteins which control microtubule dynamics (Fig. 4).[37-39] The intracellular tyrosine kinase Fyn was found to bind to the intracellular domain of the plexin-A2 receptor upon stimulation with Sema3A, and to phosphorylate it. Activated Fyn also phosphorylates the serine-threonine kinase Cdk-5 which also associates with plexin-A2 in response to Sema3A. One of the Cdk5 substrates is Tau, a protein that functions as a regulator of microtubule dynamics, indicating that recruitment of Fyn by activated plexin-A2 is part of a mechanism by which activated type-A plexins regulate the organization of the cytoskeleton.[40]

Lastly, the intracellular domain of the drosophila homologue of plexin-A1, plexin-A, contains a binding site that enables association with the flavoprotein oxidoreductase MICAL (molecule interacting with CasL), which was found to be essential for correct semaphorin-1a induced axon repulsion in *Drosophila*. MICAL has several vertebrate homologues which have been found to be important for the transduction of Sema3A and Sema3F signals, although the exact mechanism remains to be elucidated.[41,42]

The Role of the Neuropilins in VEGF Signaling

Vascular endothelial growth factor (VEGF) (also known as VEGF-A) is considered to be a major angiogenic factor that plays an essential role in embryonic vasculogenesis and angiogenesis as well as in tumor angiogenesis.[43] Multiple forms of VEGF are produced as a result of alternative splicing, but three of these forms, $VEGF_{121}$, $VEGF_{165}$, and $VEGF_{189}$ are considered to be the major forms that are most frequently encountered. All VEGF forms bind and activate the VEGFR-2 tyrosine kinase receptor which seems to be essential for the transmission of VEGF-induced angiogenic signals.[44,45] The VEGFR-1 tyrosine kinase receptor[46] is also required for developmental angiogenesis.[47] However, the role of VEGFR-1 in developmental angiogenesis is considered to be primarily an inhibitory role because the intra-cellular part of VEGFR-1 is not required for correct vascular development.[48] Nevertheless, recent experiments employing VEGFR-1 function blocking antibodies indicate that contrary to previous assumptions, this receptor plays an important active role in VEGF induced angiogenesis, presumably as a recruiter of bone marrow derived precursor cells to sites of active angiogenesis.[49]

The major VEGF splice forms differ with respect to the expression of exons 6 and 7 of the VEGF gene. Exons 6 and 7 encode independent heparin binding domains that are incorporated into the longer VEGF forms. In contrast, the shortest VEGF splice form, $VEGF_{121}$, lacks exons 6 and 7 altogether and does not bind to heparin. Experiments designed to characterize differences between the VEGF splice forms lead to the identification of splice form specific VEGF receptors in endothelial cells.[50] This receptor turned out to be np1.[51] Subsequently, it was found that the np2 also functions as a splice form specific VEGF receptor.[52] The heparin binding domains contained in exons 6 and 7 of the VEGF gene enable VEGF forms that contain these exons to bind to neuropilins. $VEGF_{121}$ does not bind to neuropilins while $VEGF_{165}$, which contains the peptide encoded by exon 7 binds to np1 and to np2 and $VEGF_{145}$ binds to np2 but not to np1.[51,52] The VEGF family contains four additional members. These are PlGF and VEGF-B which bind to VEGFR-1 but not to VEGFR-2. The heparin binding form of PlGF as well as VEGF-B were also found to bind to np1.[53,54] The last VEGF family members are the lymphangiogenesis promoting agents VEGF-C and VEGF-D.[55] VEGF-C and VEGF-D bind to the VEGFR-2 and to the VEGFR-3 receptor, which are primarily expressed on lymphatic endothelial cells, enabling them to induce angiogenesis as well as lymphangiogenesis.[56-58] Interestingly, it was observed that VEGF-C also binds to neuropilin-2, which in turn is highly expressed in lymphatics.[59,60] However, whether np2 transduces VEGF-C signals is still unclear.

Since both neuropilins function as splice form specific VEGF receptors, it was not surprising that they were found to affect VEGF signaling and function in various experimental systems. The binding of $VEGF_{165}$ to np1 enhances $VEGF_{165}$ induced migration of endothelial cells in cells that express in addition to np1 the VEGF receptor VEGFR-2.[51] Mice lacking functional np1 receptors suffer from impaired neural vascularization and from defects in the development of large arteries such as branchial arch arteries. In addition, the development of the heart was strongly impaired in these mice, and failure of heart function was responsible for their premature death.[61] In contrast, the role of np2 in VEGF induced vasculogenesis and angiogenesis is less clear. The vasculature of mice lacking a functional np2 receptor develops normally except for defects observed at birth in small lymphatic vessels.[60,62] However, these mice do not respond to $VEGF_{165}$ by retinal angiogenesis indicating that np2 is also important for angiogenesis.[63] The importance of np2 to vascular development is also highlighted in experiments in which mice lacking both functional neuropilins were generated. These mice display a total lack of endothelial cells,[64] and their phenotype therefore resembles the phenotype of mice lacking functional VEGFR-2 receptors.[65] Furthermore, mice lacking a functional np2 gene and containing only one functional np1 gene also displayed vascular abnormalities that were more severe than those observed in mice that lack both np1 alleles.[64]

Binding/competition experiments have demonstrated that the $VEGF_{165}$ and Sema3A binding domains of np1 overlap.[21] The overlap between the $VEGF_{165}$ and Sema3A binding domains in np1 means that Sema3A could modulate $VEGF_{165}$ activity by competition for binding to np1. It was indeed found that Sema3A can inhibit the angiogenic activity of $VEGF_{165}$ in in-vitro angiogenesis assays.[66] This is not true in the case of Sema3F and np2. Sema3F binding is not inhibited by $VEGF_{165}$ indicating that np2 contains separate binding sites for $VEGF_{165}$ and Sema3F.[67] Knock-in mice expressing a np1 variant lacking Sema3A binding ability but retaining VEGF binding displayed normal vascular development but abnormal neural development indicating that the VEGF binding ability of np1 is critical for proper vascular development but the semaphorin binding activity is not critical for vascular development. In contrast, the Sema3A binding ability is required in addition to the VEGF binding ability for proper heart development.[68]

The Role of Class-3 Semaphorins in the Control of Angiogenesis and Tumor Progression

Sema3A induced signaling mediated by np1 does not seem to be important for the development of the vasculature based upon the analysis of a np1 variant that binds VEGF but not Sema3A.[68] However, in other systems it was observed that class-3 semaphorins can modulate the development of the vasculature. Implantation of Sema3A containing beads in developing chick embryo forelimbs inhibits limb vascularization.[69] In another study it was found that Sema3A produced by endothelial cells modulates vascular branching in the developing chick brain by modulating integrin function.[70] However, no effects of Sema3A on tumor progression or tumor angiogenesis have been reported to date.

Even though VEGF binds efficiently to np2,[52] it does not inhibit the binding of Sema3F to np2, indicating that the semaphorin and VEGF binding domains of np2 are separate.[67] Nevertheless, in-vitro experiments have shown that Sema3F can inhibit both VEGF and bFGF induced proliferation of endothelial cells. In addition, Sema3F was able to inhibit bFGF and $VEGF_{165}$ induced angiogenesis in two in-vivo model systems, and inhibited the development of tumors from xenografted HEK-293 cells even though it did not inhibit the proliferation of the HEK-293 cells in cell culture. Furthermore, the tumors that did develop were small and contained a much reduced density of blood vessels.[71] These observations suggest that Sema3F can initiate an anti-angiogenic signaling cascade, possibly through the activation of np2/plexin signaling. In a different study it was found that Sema3F repels endothelial cells, indicating that it could inhibit angiogenesis through repulsion of angiogenic sprouts. Indeed, tumors developing from malignant melanoma cells expressing recombinant Sema3F were poorly vascularized, and the metastatic ability of cells contained in these tumors was strongly impaired. The effects on the metastatic potential of the melanoma cells were probably partially due to inhibition of angiogenesis and partially due to a direct effect on the tumor cells.[72]

Direct effects of Sema3F on the behavior of tumor cells were also noted in additional cell types. Loss of the Sema3F gene is associated with the development of small cells lung carcinoma (SCLC).[73,74] When reexpressed in SCLC cells Sema3F completely suppressed the tumorigenicity of the cells.[75] Sema3F also inhibited the attachment and spreading and repulsed MCF-7 breast cancer cells.[76,77] Sema3F binds to np2 which is considered to function, along with associated type-A plexins, as the signal transducing receptor for Sema3F.[18] However, Sema3F also binds to np1, although with a 10 fold lower affinity. Interestingly, the effects of Sema3F on the attachment and spreading of MCF-7 are mediated by the np1 receptor.[76] These effects were associated with the down regulation of E-cadherin expression.[77]

The class-3 semaphorin Sema3B was also found to function as a suppressor of small cell lung carcinoma development. Interestingly, a point mutation in Sema3B (T415I) reduces its anti-tumorigenic activity leading to increased risk of lung cancer in affected African-American and Latino-American populations.[78] Its expression in the tumorigenic cells is reduced due to either loss of heterozygosity or due to promoter hypermethylation.[79] As in the case of Sema3F, reexpression of recombinant Sema3B in SCLC cells inhibits their tumorigenic properties.[80,81] The mechanism used by Sema3B to inhibit tumor development is unclear. One report indicates that Sema3B antagonizes np1 mediated effects of $VEGF_{165}$ on the tumor cells thereby inducing apoptosis and inhibiting the proliferation of various lung and breast cancer derived tumor cells.[82] However, it was recently reported that the primary Sema3B receptor is np2.[83] Surprisingly, mice lacking a functional Sema3B gene display no gross abnormalities, indicating that it does not fulfill a nonredundant role in embryonic vascular development.[84]

Sema3C is another class-3 semaphorin that seems to signal through np2 or through np1/np2 complexes.[85] Recently, it was found that in the presence of plexin-D1, Sema3C can induce signal transduction via np1 as well as via np2.[86] The heart of mice lacking a functional Sema3C gene does not develop normally.[87] This observation fits with the phenotype of mice lacking a functional plexin-D1 gene which also suffer from heart defects and in addition from vascular patterning defects such as abnormal organization of inter-somitic blood vessels.[86,88] Thus, plexin-D1 mediated Sema3C signaling may be responsible for part of the observed defects such as heart defects. However, it was recently reported that plexin-D1 also functions as a receptor for Sema3E, the only class-3 semaphorin which does not bind to neuropilins.[13] Sema3E induces the collapse of the cytoskeleton of COS-7 cells expressing plexin-D1 and repels chick embryo blood vessels. Interestingly, mice lacking a functional Sema3E gene displayed vascular anomalies similar to the defects observed in mice lacking plexin-D1 such as abnormal patterning of intersomitic blood vessels. These abnormalities were generated independently of the presence or absence of functional neuropilins, suggesting that semaphorins that signal via neuropilins such as Sema3C do not play a role in the patterning of these vessels.[13]

The role of Sema3E in tumor progression had not yet been studied in detail. High level expression of Sema3E was observed in several metastatic cell lines originating from mouse mammary adenocarcinoma tumors, indicating that high Sema3E expression levels are linked to tumor progression.[89] More recent work revealed that over-expression of recombinant Sema3E in mammary adenocarcinoma cells induced the ability to form experimental lung metastases. In vitro, Sema3E protein enhanced the migration and proliferation of endothelial cells and pheochromocytoma cells. Interestingly, the active Sema3E form that promotes these activities is a 61 kDa Sema3E form that is generated by furin-dependent proteolytic cleavage.[90] These observations indicate that besides being the only class-3 semaphorin that does not require a neuropilin for activation of signal transduction, Sema3E also differs from the other class-3 semaphorins by being the only one that seems to promote tumor progression. Since Sema3E enhances proliferation and migration of endothelial cells, it is also likely to function as a pro-angiogenic factor although this still requires proof.

Cell Surface Attached Semaphorins as Modulators of Angiogenesis and Tumor Progression

Class 4 Semaphorins and Their Role in Tumor Progression

The membrane-anchored class 4 semaphorin Sema4D has recently become a focus of intensive research as a result of its recently discovered role in immune recognition,[91] and as a result of its newly discovered role as a regulator of tumor cell invasiveness. HGF/SF (hepatocyte growth factor/scatter factor) induces scattering, invasion, proliferation and branching morphogenesis, and plays a role as a regulator of invasiveness and tumor spread in many types of tumors.[92,93] The HGF/SF tyrosine-kinase receptor MET as well as MET like receptors such as the RON receptor for macrophage stimulating protein contain a conserved sema domain, and were recently found to associate and form complexes with several types of receptors belonging to the plexin-B subfamily. When a plexin-B1/MET complex is challenged with Sema4D, the tyrosine-kinase activity of MET is activated, just as if it were activated by HGF/SF, leading to the phosphorylation of MET. This in turn results in the stimulation of invasive growth.[94,95] These observations implicate Sema4D as a potential inducer of tumor invasiveness and tumor progression, although this has yet to be demonstrated experimentally. HGF/SF also functions as an inducer of angiogenesis,[96,97] indicating that Sema4D may also be able to induce angiogenesis via a similar mechanism. Indeed, recent evidence indicates that Sema4D induces angiogenesis as a result of its interaction with the MET receptor.[98]

The tyrosine-kinase receptor ErbB-2 which lacks a ligand of its own was also found to form complexes with plexin-B1 and to be phosphorylated in response to Sema4D.[99] Mutations in ErbB-2 are known to play a role in the induction of tumorigenesis in breast cancer as well as in other types of cancer.[100] These observations suggest that activating mutations in plexin-B1 may perhaps be able to induce activation of ErbB2 and MET in the absence of any ligand and thereby contribute to tumorigenesis. These possibilities will have to be tested in the future.

Interestingly, it was also reported that activation of plexin-B1 by Sema4D can induce angiogenesis via a MET independent mechanism. These MET independent pro-angiogenic effects were found to be dependent upon the COOH-terminal PDZ-binding motif of plexin-B1 which binds the Rho GEF proteins PDZRhoGEF and LARG. The pro-angiogenic effects of Sema4D were found in this study to be dependent upon the the activation of Rho-associated signaling pathways.[101]

Class 5 and 6 Semaphorins in Tumor Progression

Class-5 semaphorins are anchored to cell membranes and are characterized by seven type 1 thrombospondin repeats functionally important for tumorigenicity and metastasis (Fig. 1). A random p-element insertion screen was used to identify genes that modulate tumor progression and tumorigenicity. One of the genes identified in this screen was the *Drosophila* homologue of the Sema5C gene, lethal giant larvae. Deletion of the lethal giant larvae gene leads to the generation of highly invasive and widely metastatic tumors in adult flies. Further experiments indicated that lethal giant larvae probably associates, via its thrombospondin repeats, with the TGF-β like ligand DPP somehow modulating DPP induced signal transduction.[102]

Vertebrates possess a Sema5C homologue as well as additional class-5 semaphorins (Sema5A, Sema5B and Sema5D),[103,104] indicating that these homologues may play a role in the development and progression of human tumors too, although this hypothesis still requires experimental proof. Recently, this assumption was strengthened by experiments that have shown that Sema5A activates plexin-B3, a receptor that was also found to form complexes with the MET tyrosine-kinase receptor[25] and by findings indicating that Sema5A plays a role in the remodeling of the cranial vascular system.[105] This is therefore one more example of a semaphorin that interacts directly with a B type plexin and activates as a result the MET tyrosine-kinase receptor which had been previously shown to play a role in tumor progression and tumor angiogenesis.

A class-6 semaphorin that may be involved in tumor progression is Sema6B. Sema6B was found to be expressed in two different human glioblastoma cell lines, and its levels were down-regulated by trans-retinoic acid, an anti-tumorigenic, differentiation promoting agent.[106] Another class-6 semaphorin, Sema6D, may function as a pro-angiogenic factor. Sema6D binds directly to plexin-A1 which was found to associate directly with the VEGFR-2 VEGF receptor. These findings suggested that Sema6D binding to Plexin-A1 enhances the VEGFR2-signaling pathway. When the effect of exogenous Sema6D on the tyrosine phosphorylation of VEGFR2 was tested, it was found that Sema6D synergistically enhances VEGF-induced phosphorylation of VEGFR2, indicating that Sema6D and VEGF collaborate in signaling through a Plexin-A1/VEGFR2 complex, and indicating that Sema6D has the potential to enhance VEGF induced angiogenesis.[26]

Semaphorins as Direct Regulators of Tumor Cell Behavior

Many types of tumor cells express neuropilins and plexins. It was recently shown that Sema3F can also affect the behavior of tumor cells directly, via np1 and np2 receptors expressed on the tumor cells. Sema3F inhibits VEGF-induced spreading of MCF-7 breast cancer cells by inhibiting np1-mediated signaling[76] and repulses C100 breast cancer cells as a result of its binding to np2 receptors expressed by this cell type.[77] Similarly, Sema3B was found to antagonize the anti-apoptotic effects of VEGF in NCI-H1299 lung cancer derived cells, probably by interfering with neuropilin mediated VEGF signaling in these cells.[82] These findings indicate that Sema3B and Sema3F can directly affect the behavior of tumor cells expressing neuropilins.

Conclusions

The identification of the neuropilins as receptors for angiogenic factors of the VEGF family implied that the alternative neuropilin ligands belonging to the class-3 semaphorin subfamily may also regulate angiogenesis and vascular function. The past few years have indeed confirmed that some of these semaphorins are modulators of angiogenesis. Furthermore, semaphorins such as Sema4D which do not interact with neuropilins have also been found to regulate angiogenesis. Since tumor development depends on tumor angiogenesis it is not surprising that some of these semaphorins also affect tumor development. Furthermore, since neuropilins and plexins are expressed quite ubiquitously, it is no surprise that semaphorins were also found to directly affect the behavior of many types of tumor cells directly, and thus to function as promoters or inhibitors of tumor progression. We have so far only touched the tip of the iceberg as far as understanding the role of the semaphorins in tumor progression is concerned. The next few years will likely result in a flood of information, out of which it is hoped that new and efficient cancer therapies will emerge.

Acknowledgements

This work was supported by grants from the Israel Science Foundation (ISF), German-Israeli Binational Foundation (GIF), International Union against Cancer (AICR) and by the Rappaport Family Institute for Research in the Medical Sciences of the Faculty of Medicine at the Technion, Israel Institute of Technology (to G. Neufeld).

References

1. Luo Y, Raible D, Raper JA. Collapsin: A protein in brain that induces the collapse and paralysis of neuronal growth cones. Cell 1993; 75:217-227.
2. Goodman CS, Kolodkin AL, Luo Y et al. Unified nomenclature for the semaphorins collapsins. Cell 1999; 97:551-552.
3. Gherardi E, Love CA, Esnouf RM et al. The sema domain. Curr Opin Struct Biol 2004; 14:669-678.
4. Wang X, Kumanogoh A, Watanabe C et al. Functional soluble CD100/Sema4D released from activated lymphocytes: Possible role in normal and pathologic immune responses. Blood 2001; 97:3498-3504.
5. Isbister CM, O'connor TP. Mechanisms of growth cone guidance and motility in the developing grasshopper embryo. J Neurobiol 2000; 44:271-280.
6. Shirvan A, Ziv I, Fleminger G et al. Semaphorins as mediators of neuronal apoptosis. J Neurochem 1999; 73:961-971.
7. Bagnard D, Sainturet N, Meyronet D et al. Differential MAP kinases activation during Semaphorin3A-induced repulsion or apoptosis of neural progenitor cells. Mol Cell Neurosci 2004; 25:722-731.
8. Tamamaki N, Fujimori K, Nojyo Y et al. Evidence that Sema3A and Sema3F regulate the migration of GABAergic neurons in the developing neocortex. J Comp Neurol 2003; 455:238-248.

9. Bagnard D, Vaillant C, Khuth ST et al. Semaphorin 3A-vascular endothelial growth factor-165 balance mediates migration and apoptosis of neural progenitor cells by the recruitment of shared receptor. J Neurosci 2001; 21:3332-3341.
10. Brambilla E, Constantin B, Drabkin H et al. Semaphorin SEMA3F localization in malignant human lung and cell lines: A suggested role in cell adhesion and cell migration. Am J Pathol 2000; 156:939-950.
11. Barberis D, Artigiani S, Casazza A et al. Plexin signaling hampers integrin-based adhesion, leading to Rho-kinase independent cell rounding, and inhibiting lamellipodia extension and cell motility. Faseb J 2004; 18:592-594.
12. Takahashi T, Fournier A, Nakamura F et al. Plexin-neuropilin-1 complexes form functional semaphorin-3A receptors. Cell 1999; 99:59-69.
13. Gu C, Yoshida Y, Livet J et al. Semaphorin 3E and Plexin-D1 control vascular pattern independently of neuropilins. Science 2005; 307:265-268.
14. Tamagnone L, Artigiani S, Chen H et al. Plexins are a large family of receptors for transmembrane, secreted, and GPI-Anchored semaphorins in vertebrates. Cell 1999; 99:71-80.
15. Takagi S, Hirata T, Agata K et al. The A5 antigen, a candidate for the neuronal recognition molecule, has homologies to complement components and coagulation factors. Neuron 1991; 7:295-307.
16. Fujisawa H, Takagi S, HirataT. Growth-Associated expression of a membrane protein, neuropilin, in Xenopus optic nerve fibers. Dev Neurosci 1995; 17:343-349.
17. Kolodkin AL, Levengood DV, Rowe EG et al. Neuropilin is a semaphorin III receptor. Cell 1997; 90:753-762.
18. Chen H, Chedotal A, He Z et al. Neuropilin-2, a novel member of the neuropilin family, is a high affinity receptor for the semaphorins Sema E and Sema IV but not Sema III. Neuron 1997; 19:547-559.
19. Giger RJ, Urquhart ER, Gillespie SK et al. Neuropilin-2 is a receptor for semaphorin IV: Insight into the structural basis of receptor function and specificity. Neuron 1998; 21:1079-1092.
20. He Z, Tessier-Lavigne M. Neuropilin is a receptor for the axonal chemorepellent Semaphorin III. Cell 1997; 90:739-751.
21. Gu C, Limberg BJ, Whitaker GB et al. Characterization of neuropilin-1 structural features that confer binding to semaphorin 3A and vascular endothelial growth factor 165. J Biol Chem 2002; 277:18069-18076.
22. Nakamura F, Tanaka M, Takahashi T et al. Neuropilin-1 extracellular domains mediate Semaphorin D/III- induced growth cone collapse. Neuron 1998; 21:1093-1100.
23. Negishi M, Oinuma I, Katoh H. Plexins: Axon guidance and signal transduction. Cell Mol Life Sci 2005; 62(12):1363-1371.
24. Gutmann-Raviv N, Kessler O, Shraga-Heled N et al. The neuropilins and their roll in tumorigenesis and tumor progression. Cancer Lett 2006; 231:1-11.
25. Artigiani S, Conrotto P, Fazzari P et al. Plexin-B3 is a functional receptor for semaphorin 5A. Embo Rep 2004; 5:710-714.
26. Toyofuku T, Zhang H, Kumanogoh A et al. Dual roles of Sema6D in cardiac morphogenesis through region-specific association of its receptor, Plexin-A1, with off-tack and vascular endothelial growth factor receptor type 2. Genes Dev 2004; 18:435-447.
27. Takahashi T, Strittmatter SM. PlexinA1 autoinhibition by the plexin sema domain. Neuron 2001; 29:429-439.
28. Comoglio PM, Trusolino L. Invasive growth: From development to metastasis. J Clin Invest 2002; 109:857-862.
29. Oinuma I, Ishikawa Y, Katoh H et al. The Semaphorin 4D receptor Plexin-B1 is a GTPase activating protein for R-Ras. Science 2004; 305:862-865.
30. Zhang Z, Vuori K, Wang H et al. Integrin activation by R-ras. Cell 1996; 85:61-69.
31. Zanata SM, Hovatta I, Rohm B et al. Antagonistic effects of Rnd1 and RhoD GTPases regulate receptor activity in Semaphorin 3A-induced cytoskeletal collapse. J Neurosci 2002; 22:471-477.
32. Perrot V, Vazquez-Prado J, Gutkind JS. Plexin B regulates Rho through the guanine nucleotide exchange factors Leukemia-associated RhoGEF (LARG) and PDZ-RhoGEF. J Biol Chem 2002; 278:26111-26119.
33. Aurandt J, Vikis HG, Gutkind JS et al. The semaphorin receptor plexin-B1 signals through a direct interaction with the Rho-specific nucleotide exchange factor, larg. Proc Natl Acad Sci USA 2002; 99:12085-12090.
34. Hall A. Rho GTPases and the control of cell behaviour. Biochem Soc Trans 2005; 33:891-895.
35. Arimura N, Menager C, Kawano Y et al. Phosphorylation by Rho kinase regulates CRMP-2 activity in growth cones. Mol Cell Biol 2005; 25:9973-9984.

36. Harden N, Lee J, Loh HY et al. A Drosophila homolog of the Rac- And Cdc42-activated serine/threonine kinase PAK is a potential focal adhesion and focal complex protein that colocalizes with dynamic actin structures. Mol Cell Biol 1996; 16:1896-1908.
37. Mitsaki N, Inatome R, Takahashi S et al. Involvement of Fes/Fps tyrosine kinase in semaphorin3A signaling. EMBO J 2002; 21:3274-3285.
38. Gu YJ, Ihara Y. Accelerated publication - Evidence that collapsin response mediator protein-2 is involved in the dynamics of microtubules. J Biol Chem 2000; 275:17917-17920.
39. Brown M, Jacobs T, Eickholt B et al. Alpha2-chimaerin, cyclin-dependent Kinase 5/P35, and its target collapsin response mediator protein-2 are essential components in semaphorin 3A-induced growth-cone collapse. J Neurosci 2004; 24:8994-9004.
40. Sasaki Y, Cheng C, Uchida Y et al. Fyn and Cdk5 mediate Semaphorin-3A signaling, which is involved in regulation of dendrite orientation in cerebral cortex. Neuron 2002; 35:907.
41. Terman JR, Mao T, Pasterkamp RJ et al. MICALs, a family of conserved flavoprotein oxidoreductases, function in plexin-mediated axonal repulsion. Cell 2002; 109:887-900.
42. Pasterkamp RJ, Dai HN, Terman JR et al. MICAL flavoprotein monooxygenases: Expression during neural development and following spinal cord injuries in the rat. Mol Cell Neurosci 2006; 31:52-69.
43. Neufeld G, Cohen T, Gengrinovitch S et al. Vascular endothelial growth factor (VEGF) and its receptors. Faseb J 1999; 13:9-22.
44. Terman BI, Dougher-Vermazen M, Carrion ME et al. Identification of the KDR tyrosine kinase as a receptor for vascular endothelial cell growth factor. Biochem Biophys Res Commun 1992; 187:1579-1586.
45. Shibuya M. Vascular endothelial growth factor receptor-2: Its unique signaling and specific ligand, VEGF-E. Cancer Sci 2003; 94:751-756.
46. Devries C, Escobedo JA, Ueno H et al. The fms-like tyrosine kinase, a receptor for vascular endothelial growth factor. Science 1992; 255:989-991.
47. Fong GH, Rossant J, Gertsenstein M et al. Role of the Flt-1 receptor tyrosine kinase in regulating the assembly of vascular endothelium. Nature 1995; 376:66-70.
48. Hiratsuka S, Minowa O, Kuno J et al. Flt-1 lacking the tyrosine kinase domain is sufficient for normal development and angiogenesis in mice. Proc Natl Acad Sci USA 1998; 95:9349-9354.
49. Luttun A, Tjwa M, Moons L et al. Revascularization of ischemic tissues by PIGF treatment, and inhibition of tumor angiogenesis, arthritis and atherosclerosis by anti-Flt1. Nat Med 2002; 8:831-840.
50. Gitay-Goren H, Cohen T, Tessler S et al. Selective binding of VEGF121 to one of the three VEGF receptors of vascular endothelial cells. J Biol Chem 1996; 271:5519-5523.
51. Soker S, Takashima S, Miao HQ et al. Neuropilin-1 is expressed by endothelial and tumor cells as an isoform specific receptor for vascular endothelial growth factor. Cell 1998; 92:735-745.
52. Gluzman-Poltorak Z, Cohen T, Herzog Y et al. Neuropilin-2 and Neuropilin-1 are receptors for 165-amino acid long form of vascular endothelial growth factor (VEGF) and of placenta growth factor-2, but only neuropilin-2 functions as a receptor for the 145 amino acid form of VEGF. J Biol Chem 2000; 275:18040-18045.
53. Makinen T, Olofsson B, Karpanen T et al. Differential binding of vascular endothelial growth factor B splice and proteolytic isoforms to neuropilin-1. J Biol Chem 1999; 274:21217-21222.
54. Migdal M, Huppertz B, Tessler S et al. Neuropilin-1 is a placenta growth factor-2 receptor. J Biol Chem 1998; 273:22272-22278.
55. Tammela T, Enholm B, Alitalo K et al. The biology of vascular endothelial growth factors. Cardiovasc Res 2005; 65:550-563.
56. Veikkola T, Jussila L, Makinen T et al. Signaling via vascular endothelial growth factor receptor-3 is sufficient for lymphangiogenesis in transgenic mice. Embo J 2001; 20:1223-1231.
57. Kaipainen A, Korhonen J, Mustonen T et al. Expression of the fms-like tyrosine kinase 4 gene becomes restricted to lymphatic endothelium during development. Proc Natl Acad Sci USA 1995; 92:3566-3570.
58. Kukk E, Lymboussaki A, Taira S et al. VEGF-C receptor binding and pattern of expression with VEGFR-3 suggests a role in lymphatic vascular development. Development 1996; 122:3829-3837.
59. Karkkainen MJ, SaaristoA, Jussila L et al. A model for gene therapy of human hereditary lymphedema. Proc Natl Acad Sci USA 2001; 98:12677-12682.
60. Yuan L, Moyon D, Pardanaud L et al. Abnormal lymphatic vessel development in neuropilin 2 mutant mice. Development 2002; 129:4797-4806.
61. Kawasaki T, Kitsukawa T, Bekku Y et al. A requirement for neuropilin-1 in embryonic vessel formation. Development 1999; 126:4895-4902.
62. Giger RJ, Cloutier JF, Sahay A et al. Neuropilin-2 is required in vivo for selective axon guidance responses to secreted semaphorins. Neuron 2000; 25:29-41.

63. Shen J, Samul R, Zimmer J et al. Deficiency of Neuropilin 2 suppresses VEGF-induced retinal neovascularization. Mol Med 2004; 10:12-18.
64. Takashima S, Kitakaze M, Asakura M et al. Targeting of both mouse neuropilin-1 and neuropilin-2 genes severely impairs developmental yolk sac and embryonic angiogenesis. Proc Natl Acad Sci USA 2002; 99:3657-3662.
65. Shalaby F, Rossant J, Yamaguchi TP et al. Failure of blood-island formation and vasculogenesis in Flk-1- Deficient mice. Nature 1995; 376:62-66.
66. Miao HQ, Soker S, Feiner L et al. Neuropilin-1 mediates collapsin-1/semaphorin III inhibition of endothelial cell motility. Functional competition of collapsin-1 and vascular endothelial growth factor-165. J Cell Biol 1999; 146:233-242.
67. Gluzman-Poltorak Z, Cohen T, Shibuya M et al. Vascular endothelial growth factor receptor-1 and neuropilin-2 form complexes. J Biol Chem 2001; 276:18688-18694.
68. Gu C, Rodriguez ER, Reimert DV et al. Neuropilin-1 conveys semaphorin and VEGF signaling during neural and cardiovascular development. Dev Cell 2003; 5:45-57.
69. Bates D, Taylor GI, Minichiello J et al. Neurovascular congruence results from a shared patterning mechanism that utilizes Semaphorin3A and Neuropilin-1. Dev Biol 2003; 255:77-98.
70. Serini G, Valdembri D, Zanivan S et al. Class 3 semaphorins control vascular morphogenesis by inhibiting integrin function. Nature 2003; 424:391-397.
71. Kessler O, Shraga-Heled N, Lange T et al. Semaphorin-3F is an inhibitor of tumor angiogenesis. Cancer Res 2004; 64:1008-1015.
72. Bielenberg DR, Hida Y, Shimizu A et al. Semaphorin 3F, a chemorepulsant for endothelial cells, induces a poorly vascularized, encapsulated, nonmetastatic tumor phenotype. J Clin Invest 2004; 114:1260-1271.
73. Xiang RH, Hensel CH, Garcia DK et al. Isolation of the human semaphorin III/F gene (SEMA3F) at chromosome 3p21, a region deleted in lung cancer. Genomics 1996; 32:39-48.
74. Sekido Y, Bader S, Latif F et al. Human semaphorins A(V) and IV reside in the 3p21.3 small cell lung cancer deletion region and demonstrate distinct expression patterns. Proc Natl Acad Sci USA 1996; 93:4120-4125.
75. Xiang R, Davalos AR, Hensel CH et al. Semaphorin 3F gene from human 3p21.3 suppresses tumor formation in nude mice. Cancer Res 2002; 62:2637-2643.
76. Nasarre P, Constantin B, Rouhaud L et al. Semaphorin SEMA3F and VEGF have opposing effects on cell attachment and spreading. Neoplasia 2003; 5:83-92.
77. Nasarre P, Kusy S, Constantin B et al. Semaphorin SEMA3F has a repulsing activity on breast cancer cells and inhibits E-Cadherin-mediated cell adhesion. Neoplasia 2005; 7:180-189.
78. Marsit CJ, Wiencke JK, Liu M et al. The race associated allele of semaphorin 3B (SEMA3B) T415I and its role in lung cancer in African-Americans and Latino-Americans. Carcinogenesis 2005; 26:1446-1449.
79. Kuroki T, Trapasso F, Yendamuri S et al. Allelic loss on chromosome 3p21.3 and promoter hypermethylation of Semaphorin 3B in nonsmall cell lung cancer. Cancer Res 2003; 63:3352-3355.
80. Tomizawa Y, Sekido Y, Kondo M et al. Inhibition of lung cancer cell growth and induction of apoptosis after reexpression of 3p21.3 candidate tumor suppressor gene SEMA3B. Proc Natl Acad Sci USA 2001; 98:13954-13959.
81. Tse C, Xiang RH, Bracht T et al. Human Semaphorin 3B (SEMA3B) located at chromosome 3p21.3 suppresses tumor formation in an adenocarcinoma cell line. Cancer Res 2002; 62:542-546.
82. Castro-Rivera E, Ran S, Thorpe P et al. Semaphorin 3B (SEMA3B) induces apoptosis in lung and breast cancer, whereas VEGF165 antagonizes this effect. Proc Natl Acad Sci USA 2004; 101:11432-11437.
83. Julien F, Bechara A, Fiore R et al. Dual functional activity of Semaphorin 3B is required for positioning the anterior commissure. Neuron 2005; 48:63-75.
84. Van Der WL, Adams DJ, Harris LW et al. Null and conditional semaphorin 3B alleles using a flexible puroDeltatk loxP/FRT vector. Genesis 2005; 41:171-178.
85. Takahashi T, Nakamura F, Jin Z et al. Semaphorins A and E act as antagonists of neuropilin-1 and agonists of neuropilin-2 receptors. Nat Neurosci 1998; 1:487-493.
86. Gitler AD, Lu MM, Epstein JA. PlexinD1 and semaphorin signaling are required in endothelial cells for cardiovascular development. Dev Cell 2004; 7:107-116.
87. Feiner L, Webber AL, Brown CB et al. Targeted disruption of semaphorin 3C leads to persistent truncus arteriosus and aortic arch interruption. Development 2001; 128:3061-3070.
88. Torres-Vazquez J, Gitler AD, Fraser SD et al. Semaphorin-plexin signaling guides patterning of the developing vasculature. Dev Cell 2004; 7:117-123.

89. Christensen CR, Klingelhofer J, Tarabykina S et al. Transcription of a novel mouse semaphorin gene, M-semaH, correlates with the metastatic ability of mouse tumor cell lines. Cancer Res 1998; 58:1238-1244.
90. Christensen C, Ambartsumian N, Gilestro G et al. Proteolytic processing converts the repelling signal Sema3E into an inducer of invasive growth and lung metastasis. Cancer Res 2005; 65:6167-6177.
91. Kikutani H, Kumanogoh A. Semaphorins in interactions between T cells and antigen-presenting cells. Nat Rev Immunol 2003; 3:159-167.
92. Zhang YW, Vande Woude GF. HGF/SF-met signaling in the control of branching morphogenesis and invasion. J Cell Biochem 2003; 88:408-417.
93. Van DV, Taher TE, Derksen PW et al. The hepatocyte growth factor/Met pathway in development, tumorigenesis, and B-cell differentiation. Adv Cancer Res 2000; 79:39-90.
94. Giordano S, Corso S, Conrotto P et al. The Semaphorin 4D receptor controls invasive growth by coupling with met. Nat Cell Biol 2002; 4:720-724.
95. Conrotto P, Corso S, Gamberini S et al. Interplay between scatter factor receptors and B plexins controls invasive growth. Oncogene 2004; 23:5131-5137.
96. Bussolino F, Di Renzo MF, Ziche M et al. Hepatocyte growth factor is a potent angiogenic factor which stimulates endothelial cell motility and growth. J Cell Biol 1992; 119:629-641.
97. Grant DS, Kleinman HK, Goldberg ID et al. Scatter factor induces blood vessel formation in vivo. Proc Natl Acad Sci USA 1993; 90:1937-1941.
98. Conrotto P, Valdembri D, Corso S et al. Sema4D induces angiogenesis through met recruitment by Plexin B1. Blood 2005; 105:4321-4329.
99. Swiercz JM, Kuner R, Offermanns S. Plexin-B1/RhoGEF-mediated RhoA activation involves the receptor tyrosine kinase ErbB-2. J Cell Biol 2004; 165:869-880.
100. Peles E, Yarden Y. Neu and its ligands: From an oncogene to neural factors. Bioessays 1993; 15:815-824.
101. Basile JR, Barac A, Zhu T et al. Class IV semaphorins promote angiogenesis by stimulating Rho-initiated pathways through plexin-B. Cancer Res 2004; 64:5212-5224.
102. Woodhouse EC, Fisher A, Bandle RW et al. Drosophila screening model for metastasis: Semaphorin 5c is required for l(2)gl cancer phenotype. Proc Natl Acad Sci USA 2003; 100:11463-11468.
103. Inagaki S, Furuyama T, Iwahashi Y. Identification of a member of mouse semaphorin family. Febs Lett 1995; 370:269-272.
104. Furuyama T, Inagaki S, Kosugi A et al. Identification of a novel transmembrane semaphorin expressed on lymphocytes. J Biol Chem 1996; 271:33376-33381.
105. Fiore R, Rahim B, Christoffels VM et al. Inactivation of the sema5a gene results in embryonic lethality and defective remodeling of the cranial vascular system. Mol Cell Biol 2005; 25:2310-2319.
106. Correa RG, Sasahara RM, Bengtson MH et al. Human semaphorin 6B [(HSA)SEMA6B], a novel human class 6 semaphorin gene: Alternative splicing and all-trans-retinoic acid-dependent downregulation in glioblastoma cell lines. Genomics 2001; 73:343-348.

CHAPTER 11

Semaphorin Signaling in the Immune System

Vincent Potiron, Patrick Nasarre, Joëlle Roche, Cynthia Healy
and Laurence Boumsell*

Abstract

Semaphorins, a family of genes encoding guidance molecules in the nervous system, influence a variety of cellular mechanisms including migration, proliferation and cytoskeleton reorganization. Interestingly, many members are expressed throughout lymphoid tissues and by different immune cells like lymphocytes, NK, monocytes and dendritic cells. Besides, the array of functions semaphorins usually regulate during organogenesis coincide with several key events required for the initiation as well as the regulation of the host immune response. Thus, it is not surprising if a substantial number of them modulates immune processes such as the establishment of the immunological synapse, differentiation to effector and helper cells, clonal expansion, migration and phagocytosis. For this purpose, immune semaphorins can signal via their canonical plexin receptors but also possibly by unique discrete cell surface proteins or associations thereof expressed by, and critical to, leukocytes. A growing list of semaphorins, receptors or related molecules keep being reported in the immune system, and display nonredundant roles at controlling its integrity and efficacy.

Introduction

The immune system is a finely tuned process designed to recognize self-modified or foreign antigens, while being made tolerant to self antigen through various central or peripheral processes.

In the periphery, after its encounter with antigen, the antigen-presenting cell (APC) migrates via the lymphatic or the vascular networks towards lymphoid organ to present antigen to specific T cells. This migration step involves changes in expression and regulation of adhesion molecules and, in addition, cytoskeleton reorganization to allow crossing through endothelium barrier. This intense phase of cell-to-cell communication is achieved by the mean of secreted molecules and the formation of the so-called immunological synapse for direct contact with engagement of receptor complexes and costimulatory molecules, while other usually larger molecules such as CD45 are excluded from the synapse.

Strikingly, many members of the semaphorin family control integrin-mediated migration as well as the actin and tubulin networks dynamic in various circumstances. As semaphorins can be membrane-associated or soluble, they are thereby suitable for either direct cell-cell interactions or chimiotactism/chemorepulsion. Originally, semaphorins were described in the nervous system, where they guide axons and neuronal progenitors to their appropriate targets. The expression of semaphorins and their receptors at the surface of many lymphoid cells highlight their importance not only as neuronal modulators but also as regulators of immunity through overlapping mechanisms.

*Corresponding Author: Laurence Boumsell, INSERM U659, Faculté de Médecine, 8 rue du Général Sarrail, 94010 Créteil, France, Email: boumsell@im3.inserm.fr

Semaphorins: Receptor and Intracellular Signaling Mechanisms, edited by R. Jeroen Pasterkamp.
©2007 Landes Bioscience and Springer Science+Business Media.

CD100/*SEMA4D*

Among all semaphorins expressed in the immune system, CD100, the product of the SEMA4D gene, has been the first and the most extensively studied member (For reviews see ref. 1-4). In 1992, the cell surface protein was identified with monoclonal antibodies as a 150 kDa homodimeric molecule whose surface expression was upregulated after activation of T-lymphocytes.[5] It was named CD100 in 1993 during the 5th International Workshop and Conference on Human Differentiation Antigens, while in 1996, *SEMA4D* gene was cloned, and matched to the semaphorin and immunoglobulin (Ig) gene families. The same year, a highly similar murine homolog gene was also reported. Later, CD100 encoding gene was named SEMA4D and classified in the class IV of the semaphorin genes with regard to its structural features and membrane anchorage (see Table 1) for nomenclature of the genes and the molecules as approved by the respective Nomenclature Committees of the Human Genome, HGNC http://www.gene.ucl.ac.uk/nomenclature , and of the International Union of Immunological societies, IUIS, http://www.iuisonline.org).

Structure

CD100 is mainly found at the cell surface, as a 300 kDa disulfide-linked transmembrane homodimer (Fig. 1). Each 150 kDa unit bears a 500-amino-acid cysteine-rich domain in its NH_2-terminal region, the "sema" domain which is typical of the entire semaphorin family with degrees of sequence variation. In addition, *SEMA4D* possesses an Ig domain, a hydrophobic transmembrane segment and a cytoplasmic tail.

However, from the full-length CD100 is derived a shorter cleaved form, referred to as soluble CD100 (or sCD100); therefore CD100 is also released like a secreted class III semaphorin. The proteolytic generation of this truncated CD100 is metalloproteinase-dependent. It was shown by L.F. Brass et al, personal communication, that MMP17/TACE may be responsible for CD100 release from platelets, although the specific enzymes releasing it from activated lymphocytes has yet to be described. The cleavage site is located close to the membrane in the extracellular part of the protein, thus the 120 kDa processed monomer retains its cysteine 674 and remains dimerized by a disulfide bond. To date, there is no data indicating that soluble and transmembrane CD100 have opposite functions, while we will discuss below whether cell surface CD100 can act both as a receptor and a ligand.

Table 1. Official nomenclatures for semaphorin gene, proteins and their receptors

Semaphorin	Class	Gene Name HGNC	Molecule Name IUIS	Old Name
	III	SEMA3A		Collapsin-1, H-Sema III
	IV	SEMA4A		M-SemB
		SEMA4D	CD100	Collapsin-4, M-Sema G
	VII	SEMA7A	CD108	H-Sema K1
	Viral	SEMAVA		A39R
		SEMAVB		AHV sema
Receptor		plexin-C1	CD232 Lyb-2, CD72	VESP

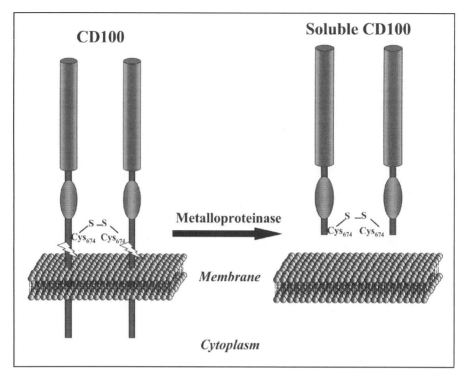

Figure 1. The membrane-associated CD100 molecule consists of a standard 500-amino-acid sema domain (cylinder) containing numerous cysteine residues, an Ig domain of the C2 type (ovoid part), a hydrophobic transmembrane region, and a cytoplasmic tail. CD100 can also undergo proteolytic processing due to a yet unidentified metalloproteinase in lymphoid cells. The resulting Soluble SEMA4D (sCD100) consists of two disulfide-linked 120-kD subunits. The cleavage site (white flash of lightning) is located COOH-terminal to the unique cysteine residue responsible for the dimerization (Cys^{674}) in the basic region that separates the Ig and transmembrane domains.

Expression

Cell surface expression of CD100 has been reported in most hematopoietic cell types. It is mainly produced by T cells where its amount is increased after their activation, while the proportion of sCD100 goes up after stimulation Nevertheless, CD100 mRNA can also be detected in nonhematopoietic tissues such as the brain, kidney and heart. CD100 is not found in the liver, the pancreas or the colon, nor in some non hematopoietic malignant cell lines.

Physiological Function

Human SEMA4D and mouse Sema4D genes have 88% homology but have not always demonstrated identical tissue distribution and effects in both species. Because of this, we will discuss separately results obtained in the human and the murine system.

Cytoplasmic Reverse Signaling and Inside-Out Regulation

The full-length CD100 has been associated in T and B-cells with a tyrosine phosphatase activity. The protein tyrosine phosphatase (PTP) was shown to be CD45, a critical receptor-like PTP involved in the differentiation of T lymphocytes and the activation of the LCK and Fyn kinases. Interestingly, CD45 plays a major role in the activation of T and B lymphocytes, while in T cells it stays outside the immunological synapse. Thus CD45 might contribute to

CD100-induced stimulation of T or B-cells by intracellular reverse signaling (Fig. 2). In this situation CD100 behaves like a receptor. In addition, this interaction appears to be regulated in the human B cell lineage according to the maturation state of the lymphocyte as CD45 could no longer be coimmunoprecipitated with CD100 in terminally differentiated B-cells.

Another way of regulating CD100 activity from the inside might be phosphorylation. CD100 coprecipitates with a serine kinase in its cytoplasmic domain, and blockade of serine phosphorylation by staurosporine enhances the shedding off the membrane. Independently, dimerization enhances cleavage of the soluble form. Thus serine phosphorylation could either negatively alter CD100 ability to be dimerized or alternatively directly inhibit proteolysis.

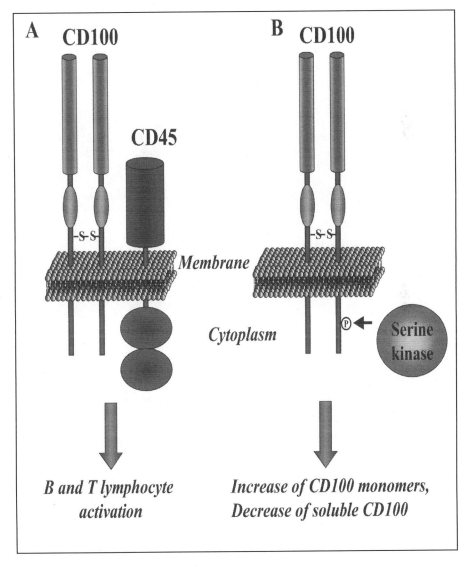

Figure 2. CD100 association with CD45 in T and B lymphocytes (A) correlates with the activation of both types of lymphocytes. CD100 can also associate with a Serine kinase (B) which could regulate the shedding process of CD100 itself.

CD100 Signaling through its Various Receptors

Signaling Via Murine CD72

To transduce cytoplasmic events, semaphorins (whether they are secreted or not) more often bind to specific membrane receptors: plexins or neuropilins., eventually associated in a complex In nonhematopoietic tissues, plexin B1 is the canonical CD100 receptor. However, plexin B1 is absent in lymphocytes, and so could not account for the binding of soluble CD100 at the surface of these cells. To address this issue, Kumanogoh and colleagues[4,6] used a substractive cDNA cloning strategy that identified CD72 as a low-affinity receptor. CD72 (or Lyb-2) is a 45 kDa type II transmembrane protein of the C-type lectin family (for a review see ref. 4,6). It is expressed in most B-cells, DCs, macrophages and subsets of T-cells, thereby supporting its involvement in CD100 signaling in leukocytes. CD72 contains two immunoreceptor tyrosine-based inhibition motifs (ITIM) in its cytoplasmic tail (V/IxYxxL/V) which become dephosphorylated after SEMA4D binding. These ITIM domains are common among inhibitory immunocoreceptors and serve to regulate B-cell-fate as a function of other signals such as B-cell receptor (BCR) agonists. In the absence of CD100, CD72 becomes heavily phosphorylated following engagement of the BCR (Fig. 3). In consequence, there is increased association with the Src-homology domain 2 (SH2)-containing protein tyrosine phosphatase-1 (SHP-1), which in turn downmodulates BCR effectors. In contrast, inhibition of CD72 with antibodies promotes survival of BCR-activated B-cells. Similarly, by binding to CD72, CD100 releases the association with SHP-1, thereby relieving the CD72 negative effect, and by this way exerts a costimulatory effect on B-cells and DCs (Fig. 3).Consequently in aged CD100-deficient mice, there is accumulation of marginal zone B cells, and they develop high auto-antibody levels and autoimmunity. Altogether other inhibitory receptors with ITIM motif, such as Fc γRIIb or CD22, and CD72 interacting with CD100, are critical in preventing excessive antibody responses.

Signaling from Human Plexin-B1 and C1

Unlike lymphocytes, monocytes and immature DCs do express detectable levels of plexins, respectively C1 and B1. Exposure to sCD100 abrogates spontaneous or MCP3 induced migration of these two cell types through their unique receptor or receptor complex and modulates cytokine production[7] (Fig. 4). This is likely to be mediated by inactivation of R-Ras, a small guanosine triphosphate (GTP)-ase that stimulates integrin-dependent adhesion. Of note, while monocytes differentiate to immature DCs in vitro and switch from plexin-C1 to plexin-B1, sCD100 keeps inducing identical activities.[7] One can then hypothesize that both receptors might activate closely related (if not the same) transduction cascades in these conditions. Further, cell surface CD100, upon interaction with plexin-B1 from the micro-environment, stimulate malignant B-CLL CD5$^+$ cell proliferation.[8] It is noteworthy that in immune cells, contrarily to soluble CD100-plexin B1 signaling in epithelial cells, cell surface CD100-plexin B1 signaling, does not involve MET or RON.

Signaling with Anti-CD100 Monoclonal Antibodies in Human Lymphoid Cells

Whether the CD100 expressed by immune cells operates as a transmembrane receptor is a matter of debate. However studies with mAbs recognizing discrete epitopes of CD100 suggest that it might act as a receptor in human lymphocytes. For example, the anti-CD100 mAb BB18, induces CD100 internalization, while inducing T cell proliferation in the presence of autologous APC and phorbol ester. In contrast, the neutralizing anti-CD100 mAb, BD16, induces the release of sCD100 from the T cell surface and the release of CD100 from associated molecules. It can also costimulate T cells with signals delivered through the TcR or the CD2 molecule. As already mentionned CD100 is associated with CD45 on both the T and the B cell surface. One fonctional consequence of this association is CD45-dependant increased T cell homotypic association in the presence of any of the 2 anti-CD100 mAb.(For reviews see ref. 1-3).

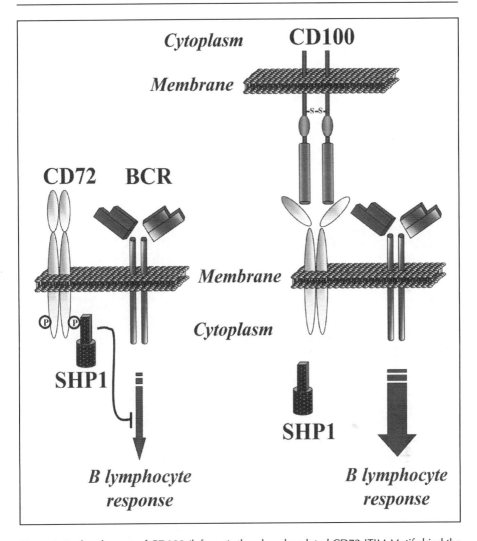

Figure 3. In the absence of CD100 (left part), the phosphorylated CD72 ITIM Motifs bind the SHP-1 protein. In this case, by inhibiting BCR effectors, SHP-1 downregulates the B-lymphocyte response induced by BCR agonists. By its binding to CD72, SEMA4D (right part) induces a release of SHP-1 from the ITIMs and increases the B-cell response.

SEMA4A

Structure, Expression and Physiological Function

Part of the class IV semaphorins, SEMA4A belongs to the same subfamily than CD100 (Table 1). It presents a similar structure and also exists as a dimer at the cell membrane. It is widely expressed in many tissues including the brain, lung, kidney, spleen and testis. In the immune system, SEMA4A is mostly found on DCs and at low level on activated B-cells, although resting B-cells and activated T-cells express marginal levels (For review see ref. 9).

The vast majority of data concerning SEMA4A comes from the murine Sema4A. Sema4A acts as a costimulator of T-cells, raise their production of interleukin (IL)-2 and proliferation in

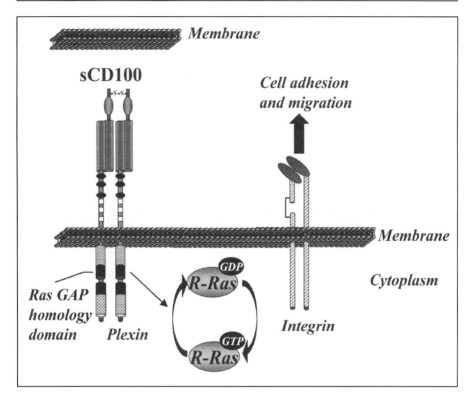

Figure 4. sCD100 binding to Plexin B1 or C1 at immature DCs or monocyte surface, respectively, could induce an inhibition of the (GTP)-ase R-Ras and a subsequent loss of cell migration capacities. This last event could be due to the role of R-Ras in the regulation of the integrin-mediated adhesion.

response to CD3 antibodies. Accordingly, Sema4A antibody inhibits allogeneic mixed leukocyte reaction (MLR), which supports a role in APC-dependent T-cell activation. Accordingly DC and T cells from knockout mice display poor allostimulatory activities and T helper differentiation.[9] Furthermore, exogenous stimulation with a Sema4A-Fc fusion protein improves the maturation of T helper lymphocytes (Th). Finally, as anti-Sema4A antibodies in animals constrain their capacity to respond in an auto-immune assay only when injected during the very first days, Sema4A might contribute to the initiation phase.

Signaling from Murine Tim-2

To investigate the function of Sema4A, a recombinant soluble form of the molecule was engineered (Sema4A-Fc) and shown to remain effective. In addition, Sema4A-Fc could bind to concanavalin A stimulated T-cells but not resting T and B-cells nor DCs. This indicated that activated T-cells specifically express a membrane receptor for Sema4A. In order to identify this Sema4A-binding protein, H. Kikutani and his team screened a mouse cDNA library. They used a strategy aimed at sorting cDNA transfected-COS7 cells based on their ability to retain biotinylated Sema4A-Fc at their surface. By doing so, cDNA clones reported the sequence for Tim-2, a member of the T-cell, Ig and mucin domain protein family (for a review on the Tim family see ref. 10). Of importance, Tim-2 had been previously characterized from a cDNA library of activated T-cells. This particularity of Tim-2 to be uniquely upregulated in activated

T-cells emphasizes its relevance to Sema4A signaling. Yet, among subpopulations, Tim-2 is nearly exclusively found in Th2. Though, even if Tim-2 knock-out mice resembles that of Sema4A in the way that both display reduced T helper immune responses, Th1, but not Th2, were demonstrated to be affected in Sema4A-deficient mice. One explanation could be a reverse signaling from Sema4A cytoplasmic domain or the presence of another receptor on Th1. Mouse Tim-2 shares the greatest identity with Tim-1, which is indeed present on Th1 and makes it a good candidate. Regrettably, while Tim-3 was excluded from Sema4A signaling, Tim-1 was not tested.

Unlike CD72, Tim-2 is not expected to be an inhibitory receptor. Instead, its cytoplasmic tail exhibits a target motif for tyrosine kinase (RTRCENQVY). Shortly after Sema4A is incubated onto cells, Tim-2 becomes phosphorylated (Fig. 5). Thus, Sema4A generates positive signals from this receptor. Unfortunately, Tim-2 precise function in T-cell stimulation or maturation remains to be elucidated. Recent findings propose that Tim-2 serves as a ferritin internalizer. Whether this function is true in vivo and has purpose in Sema4A role is ambiguous. Of note their is no human homolog for Tim2.

The expression pattern and timing of the duo Sema4A-Tim-2 during T-cell differentiation supports a role in T-cell activation by professionnal APCs (Fig. 5). In this model, APCs might preactivate T-cells, probably by presenting an antigen, which in turn induces Tim-2 expression. The binding of Sema4A from the DCs might constitute the second early step of T-cell priming. To date, there is no data concerning the intracellular pathways downstream of Tim-2 which are important in this process. On the other hand, Sema4A expression by T-cells themselves is critical to their differentiation in Th1 under skewing conditions in vitro. It appears then that another mechanism might be associated with this phenomena given the lack of Tim-2 on Th1.

Viral Semaphorins and *SEMA7A* / CD108

A39R and AHV Sema

To date, there has been two viral semaphorins described, which are secreted and share a low-complexity structure represented by a single sema domain. A39R, with 441 amino acids, is the smallest semaphorin, and is encoded by a poxvirus family member. Originally, an open reading frame (ORF) was identified from a database search in the genome of the lytic vaccinia virus, and named A39R[11] Unlike common semaphorins, A39R is not known to dimerize. The other viral semaphorin, AHV SEMA, comes from alcelaphine herpesvirus (AHV) (Table 1).

Both A39R and AHV SEMA bind to plexin-C1/ CD232, a virus encoded semaphorin protein receptor (VESPR) expressed by DCs and neutrophils. A39R raises the production by monocytes of the proinflammtory cytokines IL-6 and IL-8, as well as tumor necrosis factor alpha (TNF-α) and the level of expression of the adhesion molecule CD54.[11] In addition, A39R inhibits migration of DCs in a plexin-C1-dependent manner, through downregulation of integrin signaling as demonstrated by reduced focal adhesion kinase (FAK) phosphorylation. This phenomena was accompanied by actin cytoskeleton rearrangement. Furthermore, since phagocytosis depends on integrin function and cytoskeleton integrity, DCs and neutrophils exposed to A39R have impaired phagocytosis abilities.[12] So far, no precise role has been described for AHV SEMA. Because of its homology to SEMA7A (46%), this viral semaphorin might be used to mimick certain SEMA7A effects so as to evade the host immune response.

SEMA7A / *CD108*

SEMA7A gene was first cloned under the name CDw108, also known as the John-Milton-Hagen human blood group antigen. SEMA7A is about 80 kDa and is composed of common semaphorin features including the sema domain and an Ig domain. However, it is unique with regard to its glycosylphosphatidylinositol (GPI)-membrane anchorage. SEMA7A is expressed in activated PBMCs, in the spleen, thymus, testis, placenta and brain.[3]

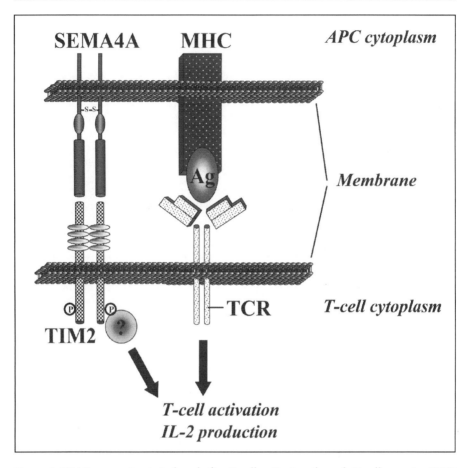

Figure 5. TIM-2 expression is induced after T-cell activation through T-cell receptor (TCR) binding to the antigen (Ag). SEMA4A binding to TIM-2 leads to the phosphorylation of TIM-2 tyrosines through a still unknown signaling. Both TCR and TIM-2 stimulations participate to T-cell activation and Interleukin-2 (IL-2) production.

Mouse DCs express plexin-C1 which binds viral semaphorins as well as SEMA7A. Despite the existence of known ligands, plexin-C1 signaling remains elusive. In in vitro experiments, it was suggested to bind plexin-C1. Lately, plexin-C1 importance for DCs was assessed by knock-out in mouse.[12] DCs from plexin-C1$^{-/-}$ mice had a migration to the lymph nodes only slightly impaired. The main consequence observed was that, in vivo, plexin-C1$^{-/-}$ DCs induced moderately reduced T-cell response. Interestingly, authors reported the presence of a GPI-linked ligand of plexin-C1 at the surface of CD40 ligand-activated DCs, which argues in favor of CD108 binding to this plexin (Fig. 6).

Like A39R, CD108 stimulates the release of IL-6, IL-8 and TNF-α by monocytes and perhaps inactivates R-Ras (Fig. 6). Nevertheless, SEMA7A promotes monocyte chemotaxis and favorizes T-cell response. Thus, SEMA7A signaling might differ at least slightly from its viral homologs.

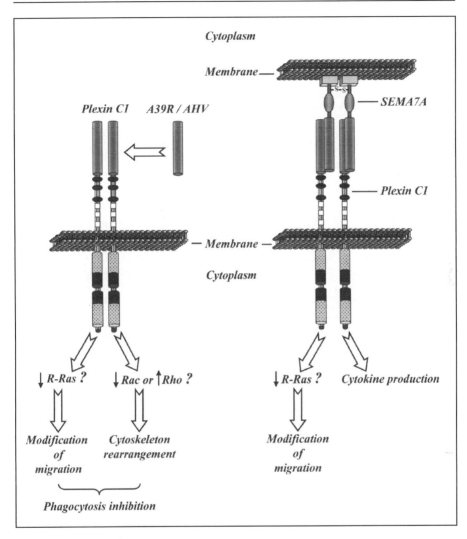

Figure 6. By its binding to Plexin C1, the viral Semaphorin A39R (left part) inhibits DCs and neutrophils phagocytosis abilities via a mechanism suggesting an inhibition of R-Ras and cytoskeleton rearrangments. SEMA7A/CD108, a GPI anchored semaphorin, is also the ligand of Plexin C1 in vitro and could induce the inhibition of R-Ras and the release of proinflammatory cytokines by monocytes.

Other Immune Semaphorins and Related Proteins

SEMA3A, SEMA3C, SEMA4B

After SEMA4D, 4A and 7A gene products were shown to play a major role in the immune system, other semaphorins became of interest in this field and were investigated. For instance, SEMA3C is upregulated in synoviocytes of rheumatoid arthritis patients, even though the biological meaning is unknown. SEMA4B is found on resting B-cells, but once again no function has yet been described. In contrast, one semaphorin -SEMA3A- indeed exhibits a role since it inhibits monocyte migration in an in vitro experiment.[1-3] Nevertheless, it is likely that given the array of functions semaphorins can influence, more might work at regulating the immune system in vivo.

Plexin-A1

There has been additional evidence for the contribution of semaphorins to immunity. One example is the distribution of their plexin receptors. DCs express plexin-A1, a receptor for SEMA3A, SEMA3F and SEMA6D. Furthermore, *plexin-A1* gene is the target of the transcription factor CIITA (the major histocompatibility complex (MHC)- class II transactivator), which is a master coactivator of MHC class II genes. In CIITA-deficient mice, plexin-A1 is not detectable on DCs. Downregulation of this receptor in a wild-type background by short hairpin RNA greatly reduces T-cell stimulation by antigen-activated DCs in vitro and in vivo. Taken together, these data strongly point to a role for plexin-A1 in T-cell-DC interaction. Whether specific semaphorins influence plexin-A1-mediated effects will be addressed in future studies.

Neuropilin-1

Similarly, another semaphorin receptor, neuropilin-1 (NRP-1), is essential for the initiation of the primary immune response. Originally, NRP1 was a heterophilic adhesion molecule identified in the nervous system . In fact, NRP-1 is also expressed on DCs and resting T-cells and is critical to the establishment of an immunological synapse between DCs and naive T-cells.[13] Functionnally, NRP-1 promotes the formation of cell clustering and its distribution becomes polarized on T-cells after the initial contact with DCs, implying that a cytoplasmic signal might be triggered. Blocking NRP-1 with antibodies substantially decreases DCs-dependent proliferation of resting T-cells. Nonetheless, NRP-1 is not involved in stimulation of activated T-cells by DCs. Altogether, these data indicate a role for NRP-1 in the initial interaction of T-cells with APCs.

Consistently, a separate study spotted NRP-1 as a possible marker for $CD4^+CD25^+$ T-cells (known as regulatory T-cells -T_{reg}-). NRP-1 was shown to be constitutively expressed on these cells independently of their activation status. In contrast, its expression was downregulated in $CD4^+CD25^-$ T-cells after activation, confirming a role in the initial phase of T-cell activation.

Semaphorins in Lymphoid Disorders

Immunosuppression by $VEGF_{165}$

NRP-1, in addition of being a receptor for some class-III semaphorins, can interact with the 165-amino acid isoform of the vascular endothelial growth factor ($VEGF_{165}$). For this reason, it appears plausible that tumor cell-derived $VEGF_{165}$ may impinge on cancer immunosurveillance through NRP-1. It is hence intriguing that $VEGF_{165}$ inhibits the development of DCs and T-cells. In view of the ability of a few semaphorins to compete with $VEGF_{165}$ for binding to NRP-1 and as a result to modulate its effectiveness, $VEGF_{165}$ impact could be linked to semaphorin signaling. It will be informative to ask if this interplay occurs here like it does in other systems such as vascular development, or if the inhibiting properties of $VEGF_{165}$ are independent.

Semaphorin Signaling in Immune Disorders

Not only do semaphorins promote proliferation of primary lymphoid cells, but they also do so in the case of leukemia. While CD100 increases $CD5^+$ B-cells proliferation, in the meantime it leads to hyperproliferation of chronic lymphocytic leukemia (CLL) cells.[8] When $CD38^+$ CLL-B-cells encounter its ligand CD31, CD100 is upregulated in proliferating cells along with CD72 downregulation while plexin-B1 is expressed, therefore switching the balance of positive/negative pressures toward proliferation.[8] The importance of semaphorins in the development/outcome of solid tumors of many kinds has been extremely emphasized in the past few years. Likewise, one might expect them to play a decisive role in the progression of leukemic diseases.

Soluble CD100[14] has also been shown to play a direct role in vitro in the inapropriate cross talk between activated T lymphocytes and neural precursor cells expressing plexin B1. In vivo activated T cells infiltrating the Central Nervous System in interaction with neural cells, particularly oligodendrocytes, could sustain the onset and progression of demyelination and axonal degeration in neuroinflammatory diseases, such as HTLV1 associated myelopathy and MS.[14] It is relevant to mention here that neither CD100 K.O. mice nor anti-sema4A treated mice develop EAE (experimental allergic encephalomyelitis). Thus semaphorin gene product, from the class IV, such as CD100 and sema4A , would be involved in the onset and the late phase of neuroinflammatory diseases. It would be of interest to develop neutralizing antibodies to prevent/treat these diseases.

Conclusion

Semaphorins, first described as nerve guidance molecules are now known as mediators of the immune network. Despite the fact that they can use canonical well described receptors in other tissues, they can also utilize distinct receptors or receptor complexes specific of the immune system. A summary of our present knowledge in this field is presented in Table 2.

Table 2. Role of semaphorins and receptors in the immune system

Sema Class	Molecule	Cell Surface Expression	Receptor and its Expression	Function
III	SEMA3A	not reported	not NP1	inhibition of monocyte migration
IV	SEMA4A	DCs, B-cells, Th1	Tim-2 (activated T-cells, Th2)	enhances T-cell activation, production of IL-2 and priming/Th1-2 regulation,
	full-length CD100	Most hematopoietic cells, increased by T cell activation	plexin-B1 (CLL, CD5$^+$ B-cells), CD72 (B-cells)	proliferation of CLL and CD5$^+$ B-cells / T-cell priming
	soluble CD100		plexin-B1 (immature DCs), plexin-C1 ?(monocytes), CD72 (B-cells, DCs)	inhibition of monocyte and immature DC migration/ co-stimulation (with CD40) of B-cells and DCs / proliferation of CD5$^+$ B-cells
VII	CD108	monocytes, activated PBMCS, thymus	plexin-C1	monocyte stimulation/ production of the proinflammatory cytokines IL-6 and 8
Viral	A39R	vaccinia virus, mouse pox virus	plexin-C1 (DCs, neutrophils)	phagocytosis inhibition/ cytoskeleton rearrangement and integrin inhibition/ production of the proinflammatory cytokines IL-6 and 8
	AHV SEMA	alcelaphaline herpes virus	plexin-C1 (DCs, neutrophils)	upregulation of CD54
Unidentified			plexin-A1 (DCs) NP1 (DCs, resting T-cells)	T-cell-DCs interaction T-cell-DCs homophilic interaction

Description of anomalies in semaphorin signaling with regard to immune pathologies is yet in its infancy and more has to be known to understand the function of this large family of proteins. At present it is not known, may-be because of the redundant expression of semaphorins, whether semaphorins act in vivo for the immune system as attractive or repulsive clues to guide or restrict the migration of immune cell subsets to specific lymphoid organs.

Abbreviations

AHV: alcelaphine herpesvirus; APC: antigen presenting cell; BCR: B-cell receptor; CD: cluster of differentiation; CLL: chronic lymphocytic leukemia; DC: dendritic cell; FAK: focal adhesion kinase; GTP: guanosine triphosphate; Ig: immunoglobulin; ITIM: immunoreceptor tyrosine-based inhibition motif; IL: interleukin; MHC: major histocompatibility complex; NRP: neuropilin; PTP: protein tyrosine phosphatase; SH2: Src-homology domain 2; SHP-1: (SH2)-containing protein tyrosine phosphatase-1; TCR: T-cell receptor; Th: T Helper lymphocyte; TNF: tumor necrosis factor; Tim-2: T-cell immunoglobulin and mucin domain protein 2; VEGF: vascular endothelial growth factor; VESPR: virus encoded semaphorin protein receptor.

References

1. Delaire S, Elhabazi A, Bensussan A et al. CD100 is a leukocyte semaphorin. Cell Mol Life Sci 1998; 54:1265-1276.
2. Bismuth G, Boumsell L. Controlling the immune system through semaphorins. Sci STKE 2002; 128RE4.
3. Elhabazi A, Marie-Cardine A, Chabbert-de Ponnat I et al. Structure and function of the immune semaphorin CD100/SEMA4D. Critical Rev In Immunol 2003; 23:65-81.
4. Kikutani H, Kumanogoh A. Semaphorins in interactions between T cells and antigen-presenting cells. Nat Rev Immunol 2003; 3(2):159-67.
5. Bougeret C, Mansur IG, Dastot H et al. Increased surface expression of a newly identified 150-kDa dimer early after human T lymphocyte activation. J Immunol 1992; 148(2):318-23.
6. Kumanogoh A, Shikina T, Watanabe C et al. Requirement for CD100-CD72 interactions in fine-tuning of B-cell antigen receptor signaling and homeostatic maintenance of the B-cell compartment. Int Imuunol 2005; 17:1277-1282.
7. Chabbert-de Ponnat I, Marie-Cardine A, Pasterkamp RJ et al. Soluble CD100 functions on human monocytes and immature dendritic cells require plexin C1 and plexin B1, respectively. Int Immunol 2005; 17(4):439-47.
8. Deaglio S, Vaisitti T, Bergui L et al. CD38 and CD100 lead a network of surface receptors relaying positive signals for B-CLL growth and survival. Blood 2005; 105(8):3042-50.
9. Kumanogoh A, Shikina T, Suzuki K et al. Nonredundant roles of Sema4A in the immune system: Defective T cell priming and Th1/Th2 regulation in Sema4A-deficient mice. Immunity 2005; 22(3):305-16.
10. Meyers JH, Sabatos CA, Chakravarti S, Kuchroo VK. The TIM gene family regulates autoimmune and allergic diseases. Trends Mol Med 2005; 11(8):362-9.
11. Spriggs MK. Shared resources between the neural and immune systems: Semaphorins join the rank. Curr Opin Immunol 1999; 11:387-391.
12. Walzer T, Galibert L, Comeau MR et al. Plexin C1 engagement on mouse dendritic cells by viral semaphorin A39R induces actin cytoskeleton rearrangement and inhibits integrin-mediated adhesion and chemokine-induced migration. J Immunol 2005; 174(1):51-9.
13. Tordjman R, Lepelletier Y, Lemarchandel V et al. A neuronal receptor, neuropilin-1, is essential for the initiation of the primary immune response. 2002; 3(5):477-482.
14. Giraudon P, Vincent P, Vuaillat C et al. Semaphorin CD100 from activated T lymphocytes induces process extension collapse in oligodendrocytes and death of immature neural cells. J Immunol 2004; 172:1246-1255.

Index

A

α-adaptin 8
A39R 27, 90, 93, 132, 139, 140, 142
Actin 1, 6-8, 12, 16, 18, 21, 24, 25, 27, 39, 45, 46, 55, 56, 59, 60, 65, 90, 93, 95, 98, 100-102, 122, 123, 132, 139
Adult neurons 82
AHV Sema 132, 139, 142
Alzheimer's disease (AD) 4, 8
Amyloid precursor protein intracellular domain (AICD) 4, 8
Angiogenesis 32, 38, 93, 97, 98, 101, 118, 120, 123-127
Antibodies 2, 3, 5, 8, 19, 30, 53, 55, 56, 64, 67, 68, 77, 79, 84, 123, 133, 136, 138, 142, 143
Arf 13
Axon guidance 1, 5, 6, 8, 13, 32, 38, 39, 46, 47, 55, 58-60, 62, 73-78, 80-85, 90, 95, 96, 101, 103, 109, 112, 118
Axon repulsion 41, 43, 46, 123
Axonal transport 24, 25, 30, 31

B

B lymphocytes 134
Brain derived neurotrophic factor (BDNF) 27
Brevican 82

C

Ca^{2+} 53, 57
Calponin homology 41, 43, 46
Cardiac development 97, 109, 111
CasL 8, 38, 39, 41, 44, 45, 123
CD100 132-137, 142, 143
CD100/SEMA4D 133
CD108 132, 139, 140, 142
CD45 132, 134-136
CD72 39, 90, 132, 136, 139, 142
Cdk5 3-7, 12, 24, 25, 28, 30, 31, 55, 59, 123
Cell adhesion 7, 16, 19, 20, 25, 39, 43, 45, 57, 61, 62, 90, 91, 93-95, 98-100, 101-103, 115, 119
Cell adhesion molecule 7, 45, 57
Cell death 24, 25, 32
Cell migration 24, 28, 62, 90, 91, 93-99, 101-103, 115, 119, 139

Cerebellar granule cells 93
Calponin homology (CH) 43, 46
Chondroitin sulfate (CS) 58, 73, 78, 80-82, 84, 85
Chondroitinase 78-80, 82
Chordin 2
Clathrin 7, 8, 65
CNS 39, 40-43, 48, 53, 55, 82-84, 96
Cofilin 18, 25-27
Collagen gel repulsion assay 53, 55, 56
Collapsin-response-mediator protein (CRMP) 1, 2-9, 28-30, 32, 45, 65, 122, 123
COS7 cells 2, 3, 5, 64, 65, 138
Cyclin-dependent kinase 5 30, 55
Cytoskeleton 1, 5-7, 13, 21, 25, 38, 39, 43, 45-47, 55, 90, 95, 98, 100, 101, 119, 122, 123, 125, 132, 139, 140, 142

D

Dihydropyrimidinase related protein (DRP) 1
Dorsal root entry zone (DREZ) 68, 69
Dorsal root sensory ganglia (DRG) 2, 3, 8, 27, 28, 30-32, 41, 52, 55-57, 68, 69, 84, 93, 101
Drosophila 1, 8, 13, 15, 20, 38, 39, 41-43, 62, 74-76, 95, 101, 112, 123, 126

E

Endocardial cells 93, 111
Ephrin 5, 13, 30, 52, 73, 74, 78, 80, 83
Experimental allergic encephalomyelitis 143
Extracellular matrix (ECM) 45, 56, 62, 73, 74, 76, 77, 79, 83, 84, 94, 95, 97, 98, 100, 102, 109

F

F-actin 1, 7, 16, 18, 46, 65, 93
F3/contactin 64
FAD 41, 43, 46
FARP2 7, 15, 16, 19, 45, 55, 56, 59, 98, 101, 115, 116
Fibroblast growth factor (FGF) 2, 73, 81, 93
Flavoprotein monooxygenase 38, 39, 41, 46, 47, 56

G

G protein signaling 7
GABAergic neurons 93, 96
Glycosaminoglycan (GAG) 16, 73, 76, 77, 78, 82-84, 85, 132, 136
Grasshopper 52
Growth cone 1-9, 12, 13, 16, 18-20, 24-28, 30, 31, 38, 39, 43, 52, 53, 55-69, 73, 74, 76, 77, 79, 80, 84, 93, 95, 100, 103, 118, 119
Growth cone turning assay 52, 53, 57
Growth-associated binder 1 (Gab-1) 27
GSK3b 3-8, 12, 30
GTPase activating proteins (GAPs) 7, 12-15, 17, 56, 91, 98, 122
GTPases 7, 12-16, 18-21, 39, 45, 46, 56, 57, 59, 90, 98, 99, 101-103, 121, 122
Guanine nucleotide exchange factors (GEFs) 7, 12, 13, 15, 21, 56, 100, 122

H

Hedgehog 73
HEK-293 cells 124
HeLa cells 45
Heparan sulfate (HS) 73, 74, 76-78
Heparin 74-77, 80, 81, 118, 123
Heparinase 74, 77-79, 82, 83
HUVEC endothelial cells 93

I

Ig superfamily 39, 61, 62
Ig superfamily cell adhesion molecule (IgSFCAMs) 39, 61-69
Immunity 132, 142
Integrins 13, 19, 25, 39, 62, 65, 66, 90, 95, 98, 100, 103, 115

L

Lamellipodium 1, 2
Lavendustin A 31
Leucocyte migration 98
LIM domain 41, 43
LIMK1 18, 25, 26
LIMK2 25

M

MAPK 24, 31, 32, 56, 57, 66, 95, 100, 102

MICAL 1, 6, 8, 38-41, 43-48, 55, 59, 123
Microdomain signaling 56
Microtubule dynamics 5, 6, 24, 25, 28, 30, 122, 123
Monoclonal antibodies 133, 136
Monocytes 93, 132, 136, 139, 140, 142
mRNA 31, 39, 56, 59, 111, 134
Myocardial cells 93, 109, 111, 112, 114

N

Nerve growth factor (NGF) 27, 32
Netrin 31, 52, 59, 73, 74, 76-82, 84
Neural crest cells 91, 93, 96, 97, 103, 111, 115
Neurocan 82
Neuropilin 1, 12, 13, 15, 39, 41, 43, 45, 53, 55-59, 64, 68, 77, 84, 90, 91, 93, 96, 97, 100, 112, 115, 116, 118-121, 123, 127, 136, 142, 144
Neuropilin-1 1, 12, 15, 41, 53, 55-60, 77, 84, 90, 93, 96, 97, 112, 115, 116, 119, 142
Neuropilin-2 53, 55, 58, 59, 77, 84, 90, 119, 123
Nicotinamide adenine dinucleotide phosphate (NADPH) 38, 46
Noggin 2
Nrp 12, 13, 16, 19, 64, 67, 69, 142, 144
Numb 6, 8

O

Off-track 41, 43, 112
Oligodendrocytes 40, 41, 82-84, 97, 143
Organ morphogenesis 97

P

P-hydroxybenzoate hydroxylase (PHBH) 46
p21-activated kinase (PAK) 16, 18-20, 25, 27, 122
Peripheral nervous system (PNS) 2, 39, 43, 53, 55
Phosphacan 82
Phosphatidylinositol (3,4,5)-trisphosphate (PIP3) 27
Phosphatidylinositol 3-kinase 27, 28
Phospholipase D2 (PLD2) 8
PIPKIg661 16, 45, 115, 116
Plasticity 30, 31, 39, 74, 83, 95, 96, 103
Platelets 93, 133

P

PlexA 1-3, 5-8, 39, 41-43
Plexin 1, 3, 5-8, 12-21, 27, 28, 30, 38-43, 45, 46, 48, 53, 55-57, 59, 64, 65, 77, 90, 91, 93, 96-103, 111-116, 118, 120-127, 132, 136, 139, 140, 142, 143
Plexin-A1 12-17, 19, 20, 53, 55, 56, 111-116, 120-123, 126, 127, 142
Protein kinases 25, 27, 32, 39, 43, 55
Protein synthesis 31, 32, 56, 57, 59, 60
Proteoglycans 73, 74, 76-80, 82, 84, 85, 90, 100
Purkinje cells 77

R

Rab 13, 45
Rac1 6, 7, 14-21, 45, 47, 65, 98, 99, 100-102, 121
Ran 13, 45
Rapamycin 31
Reactive oxygen species (ROS) 13, 41, 45-47
Receptor internalization 57, 65, 66
Regeneration 8, 32, 74, 82-84, 97, 103
Repulsion 8, 13, 32, 41, 43, 46, 52, 53, 55-59, 62-64, 66, 68, 69, 74, 76, 77, 79, 93, 95, 100, 101, 115, 123, 124
Review 1, 12, 24, 38, 39, 74, 90, 94, 95, 97, 98, 101-103, 109, 115, 136-138
Rho 5-7, 12-16, 19-21, 25, 27, 30, 39, 47, 56, 57, 59, 99-101, 115, 122, 123, 126
Rho GTPase 115
Rho-associated kinase (ROCK) 5, 16, 18, 19, 25, 27, 101

S

Schwann cell 83
Sema 27, 39, 41-43, 46, 52, 58, 62, 90, 98, 118, 120, 121, 126, 132, 133, 135, 139, 142
  Sema3A 1-9, 12, 13, 15, 16, 18-20, 25-32, 41, 43, 45, 52, 53, 55-57, 59, 60, 62-69, 77, 80-84, 90, 94-98, 100, 101, 115, 116, 118-125, 127, 132, 141, 142
  Sema3C 52, 55, 59, 90, 97, 98, 115, 125, 141
  Sema4A 90, 98, 132, 137-139, 141-143
  SEMA4B 141
  Sema4D 12, 13, 15, 16, 19, 20, 27, 28, 32, 45, 83, 90, 91, 94, 97, 98, 102, 118, 120, 122, 123, 125-127, 134-136, 141
Sema7A 12, 32, 90, 91, 95, 100, 132, 139, 140
SEMA7A/CD108 140
Semaphorin 1, 8, 9, 12-15, 20, 24, 27, 28, 32, 38-41, 43, 46-48, 52, 53, 55-62, 64-69, 73, 74, 77, 78, 82, 84, 90, 91, 93-103, 109, 111, 112, 115, 118-120, 123-127, 132, 133, 139-144
Semaphorin gene 132, 143
Slice overlay assay 52, 53
Slingshot 26, 27
Slit 73-76, 84
Slit2 74, 75, 81, 83
Small-cell lung cancer 32
Sra1 7
SSH 26, 27
Syndecan 78

T

T lymphocytes 134, 143
Tag-1/Axonin 64
Tim-2 138, 139, 141, 144
TOAD-64 1
Trabeculation 112, 114
Transforming growth factor β (TGF-β) 73
Translation 24, 25, 31, 32, 56, 62
Tumor angiogenesis 123, 124, 126, 127
Tumor cell migration 98
Tumor progression 124-127
Tyrosine kinases 28, 30, 39, 90, 91, 98, 102

U

Ulip 1

V

Vascular endothelial growth factors (VEGF) 91, 93, 103, 112, 115, 118, 123, 124, 126, 127, 144
Vasculogenesis 115, 123
Vesicle dynamics 7

W

Wnt 73
Wortmannin 31